SNAK

The Snake Keeper's Guide

Jeffrey Protheroe & Barry Pomfret

SNAKE

The Snake Keeper's Guide

by Jeffrey Protheroe & Barry Pomfret

First published 1994 by
Air Research Publications
PO Box 223, Walton-on-Thames
Surrey, KT12 3YQ

Printed in Great Britain by
MBA Group Ltd
London N17 0HW

ISBN 1-871187- 21-4

Contents

Acknowledgements

In producing this work the authors would like to thank the following individuals, without whom this book may not have been possible.

Peter Phillips for the difficult task of producing additional sketches, Bob McCaffrey and Ian Elton for the photographs unless otherwise detailed, and Adam Wright of Coventry Reptiles for the use of his article in preparing the section on sex determination and additional photographs.

Last but not least we would like to extend our thanks to our wives, Vivette and Pat for their patience and understanding of our hobby.

Introduction

Recent changes in the types of animals sold in the average pet stores have seen a massive increase in the captive care of reptiles in general, with much focus played on the keeping of snakes.

Several good books are available through most retail outlets. However most, if not all, lack the simple 'how to..' approach which is vital for the first time or novice keeper to get started. It is also true that the majority of books tend to give detailed accounts of specialised snakes that the average keeper will never encounter, whilst the most common forms found in every pet store are often overlooked.

Thus this book has been written by two experienced herpetologists (reptile and amphibian keepers), who together take the novice through several chapters dealing with the vital questions every one need ask to successfully maintain snakes in captivity. In addition to answering the basic questions, the authors have included many of the 'tricks of the trade' to help the novice overcome problems, and thus become a responsible keeper.

Where possible this book should be purchased before the acquisition of a snake itself. Special attention should be given to the section dealing with the suitable first time purchases to help in your selection of specimens.

Both authors are founder members of the Milton Keynes Herpetological Society and have dealt with individuals problems and questions for many years. Thus the content of this book is based on working knowledge.

We sincerely hope that you will have as much enjoyment as the two of us, to the extent that perhaps one day you will be in a position to offer others help and assistance.

Characteristics of Snakes

The Scales

A snake's body is covered with different types of scales which serve much the same purpose as our skin, except that scales are very much tougher. The scales offer protection against physical hazards caused through locomotion particularly from the friction where its body meets the substrate.

Unlike fish scales, snake scales are not separate but rather an integral part of the skin. Snake scales are formed into regular patterns, varying in size and shape depending on location and function.

Ecdysis

A natural process of snakes is the periodic shedding of their skin called ecdysis or more commonly 'sloughing'. A sloughed skin looks like a partially transparent snake showing all the scales present. The frequency of the slough differs from snake to snake, ranging from twice a year in some species of viper, to several times a year in many of the pythons and colubrids (ratsnakes, kingsnakes etc). Sloughing helps to relieve the snake of old partially damaged skin, odd pieces of material trapped between scales and external parasites, whilst also allowing for growth.

The snake's skin is made up of three layers, each with a particular function. The innermost layer is the thickest containing all the pigment cells, whilst the middle layer is thin and consists of cells that are continually growing. The new cells eventually die and harden forming the external layer of scales which comprises a material called Keratin. Keratin is a fibrous sulphur containing protein which gives the scales their waterproof properties.

Periodically a secretion is produced (called exudate) which separates the older external layer from the new inner layer beneath. When this occurs the whole of the body acquires an opaque appearance which is particularly noticeable over the eyes.

During this phase it is not uncommon for a placid snake to become aggressive as its sight is impaired and it feels more vulnerable. This secretion also has a secondary function in providing a lubricant for the easy removal of the old outer layer.

The snake will remain opaque in appearance for 2 - 7 days before returning to a darker than normal colour. Once the eyes have cleared you will notice that the snake starts some peculiar actions by rubbing its snout against the substrate or against other rough surfaces.

At this stage the snake is attempting to break the old skin away from the nose and mouth area in order to start the actual process of sloughing. When the snake has succeeded in breaking the skin, it will start to unwrap the skin from its body by sliding out of it, thus turning the old skin in on itself. This is best accomplished by the snake rubbing against a tree stump or rock which tends to catch the skin in order for the snake to pull out of it.

When the snake has completely shed its skin, you will notice that it will appear very 'shiny' or translucent with its normal coloration at its brightest (this is the best time to take a photograph).

If a snake is having a problem in sloughing, one may assist it through the use of any of the following techniques:

- Placing it to soak in a shallow bowl of warm water for the night.
- Regular spraying of water leading up to and including the slough.
- Placing the snake in a warm damp cloth bag over-night.
- Placing a large bowl of warm water within the vivarium to allow the snake to soak before and during the slough.

Scientific research has shown that the onset of ecdysis leads to an increased loss of water by evaporation through and from the skin. Researchers are now of the opinion that snakes seek water during ecdysis to avoid desiccation, and not as previously thought to loosen the outer layer of skin. Therefore, one must ensure that the animal has a constant supply of fresh drinking water particularly during this period.

Temperature Requirements

All snakes are poikilothermic (literally translated as poikilos meaning various and therm meaning heat) or ectothermic which implies that the animal is dependant on external sources for bodily heat. These names give rise to the common name of 'cold-blooded' when referring to these animals. Without heat the snake will simply be incapable of catching food, digesting food, escaping enemies etc, thus heat is one of the most important aspects for any reptile if they are to survive.

Snakes in general must maintain their body temperature within the range 4°C - 38°C regardless of climatic conditions. At temperatures below 2°C - 4°C they become completely lethargic, whilst at temperatures between 38°C - 47°C (depending on the duration of the exposure), certain proteins in their tissues undergo irreversible chemical changes that can cause death.

Snakes control their body temperature through their behaviour, i.e moving into and out of shade, and consequently the habits of the animal differ considerably according to the climatic conditions. This is an important point when setting up your vivarium and will be discussed in greater detail within the section dealing with vivarium equipment and set-up. This form of behavioural thermo-regulation allows the snake to maximise the effect of their main sources of heat via solar radiation and conduction, to maintain an average body temperature within a few degrees of their preferred norm.

During aestivation, snakes are inactive from choice and are merely avoiding unfavourable conditions like the extreme temperature and lack of water; however during hibernation they are in a state of lethargy as a direct result of the low temperatures and as long as this persists they are relatively helpless.

Research both in the field and in the laboratory have shown that male snakes kept at prolonged high temperatures have reduced fertility or indeed may become infertile (Ross 1990). In addition female snakes cannot produce the ovarian follicles which will mature into eggs unless exposed to the lower temperatures along with the males.

When deciding on purchasing a snake one should investigate its temperature requirements beforehand to guarantee a comfortable habitat for the animal involved.

Psychology

Snakes are very nervous animals that are far more frightened of us than we are of them. A snake spends much of its time under logs, stones, within holes and tunnels or simply squeezed tightly into a small crevice. To make a snake feel completely at ease in captivity one must try to simulate this condition. This is best achieved by placing a hide box within the vivaria to be used to house the snake, or in natural set-ups by the provision of several hide spots through the use of artificial plants and cork bark. A basic hide box may be fabricated using cardboard cartons such as cereal packets, tissue boxes or the inner cardboard roll from toilet tissue for hatchling snakes.

Jacobson's Organ

When handling a snake, you will find that it starts to flick its tongue in and out of its mouth and wave it up and down. Some individuals still have the misconception that the snakes forked tongue carries the venom in venomous species, however, this is totally untrue.

When the snake waves its tongue about, it is simply picking up particles from the air which are then passed to a special receptive organ 'The Jacobson's Organ' which is located in the anterior part of the palate within the mouth. The Jacobson's Organ consists of a pair of structures which are lined with sensory glands that send signals to the brain along olfactory nerves. These trigger a response determined by the particles i.e. do the particles belong to a possible prey item or are they from a predator, resulting in a particular response such as attack, retreat, defend or simply ignore.

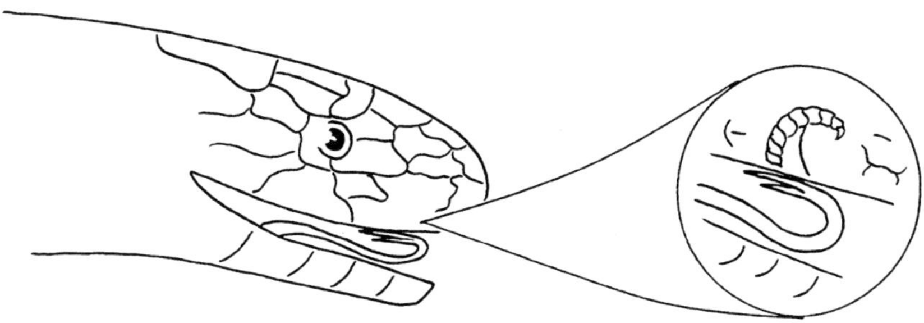

Jacobson's Organ - Note that the tongue folds back to touch the gland and pass on the particles

Eyes

Snakes have no movable eyelids, however the cornea is protected by a transparent scale called the 'brille'. Snakes have a very wide field of vision in which the slightest movement will be recorded, but which lacks sharpness, it has been shown they can only see clearly up to a few metres. However, even with this limitation in sight, snakes rely more on their sense of smell with which they can detect dead and other immobile prey such as eggs. Consequently one should exercise caution when feeding.

Ears

Snakes only possess an inner ear and a stapes (the stirrup shaped innermost ossicles of the ear) but lack any trace of an ear drum associated with it. Instead they rely on vibration to determine the distance and location of anything passing. Tests have been performed in the U.S.A. to verify this fact and found that the snakes capacity in receiving and acting on vibrations are acutely tuned. Klauber (1972) experimented with rattlesnakes and found that even though they were suspended from a pole within a bag in the centre of a room, they could detect the slightest footsteps of an individual entering the room and sound a warning rattle.

The Facial Pits

Many species of snake such as the Boids and Vipers, have an extra set of facial pits located just in front of the eyes on the lip resembling an additional set of nasal pits. These pits act as temperature differential receptors and allow the snake to determine the distance and location of prey by forming an infra-red image of the object. The pits are only useful if the prey item or target is warmer than its surroundings, thus its usefulness to the snake is improved in lower temperature zones.

Hibernation

For all of the north American colubrids described within this book (the kingsnakes, ratsnakes, garter snakes, western hognose) it is benificial to hibernate during the winter periods, for example November through until March. Some species may require shorter or even longer periods.

The importance of hibernation has already been discussed from the viewpoint of fertility, however a snake's life is prolonged if regularly hibernated every year, if even only for a short spell. In captivity many individuals state that hatchlings and yearlings do not require hibernation. This statement should be ignored, after all, in their natural habitat they would perform in exactly the same manner as the adults and generally are born within close proximity to hibernation dens. Field study has shown that hatchlings often do not have their first meal until the following spring, after their first hibernation.

In order to induce hibernation, one should lower the vivarium temperature gradually over a month (typically October) until the temperature is maintained between 7°C to 13°C (unless otherwise stated for specific subspecies). Prior to lowering the temperature, no meals should be offered for a 14 day period. Whilst the temperature is being lowered, one should monitor the animal to ensure that the last meals consumed are passed and that the animal is no longer passing faeces, only uric crystals. The animal(s) involved should be held at this temperature in a cool, dark room if possible, with access to water, but no food, until the end of February, or as detailed, with regular inspections performed in order to monitor the animal(s) condition. At the end of the required hibernation period the temperature should be raised gradually every few days over 3 - 4 weeks (typically March) until the correct spring temperature is achieved.

Locomotion

Snakes move in a number of ways depending on the species and their natural habitat.

Ground dwelling snakes, if in woodland or rocky areas, tend to move by pushing against rocks and logs or any other obstacle. This form of motion is termed 'Serpentine Locomotion' and is the result of the action of those muscles lying on either side of the backbone and attached to the sides of the vertebrae which produce a propulsive action when the surface of an object is encountered.

In areas of flatland, the snake can be seen to move by 'running' its rib-cage along the floor. This method is termed 'Rectilinear Creeping' and is the function of several hundred muscles which are attached to the rib cage and the belly areas. The contraction of one muscle lifts the belly from the ground whilst a second muscle pulls the body forward. This process is repeated from the tail through to the neck region such that the snake 'walks' along on its ventral region. This method of locomotion is used primarily in the large boids such as pythons.

Arboreal snakes move in much the same manner as ground dwellers, but use the branches available to achieve greater speeds. Some, more specialised, snakes move in a sideways direction often called 'side winding' utilised by many species living in deserts and similar environments.

Snake

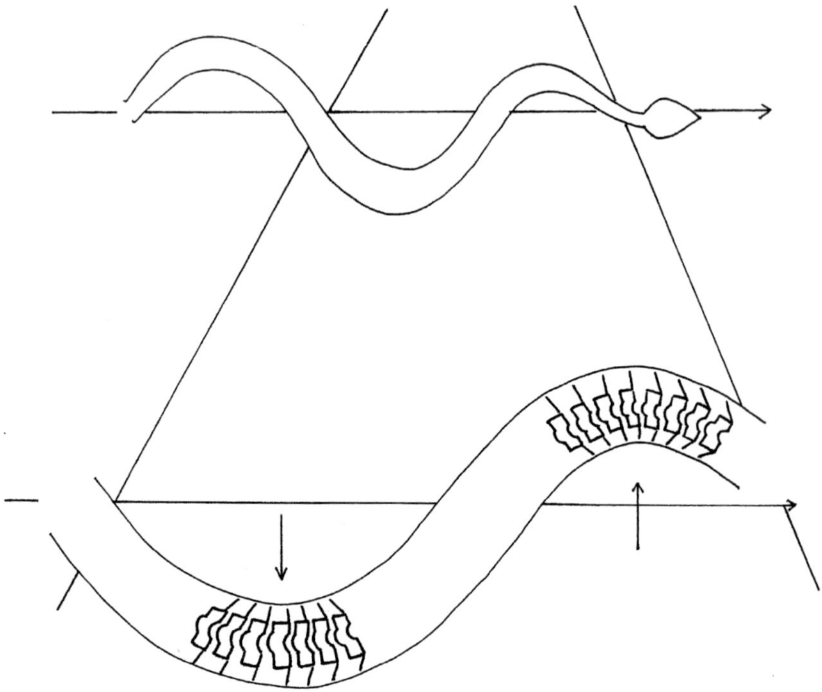

The vertebrae move to allow the snake to push forward.

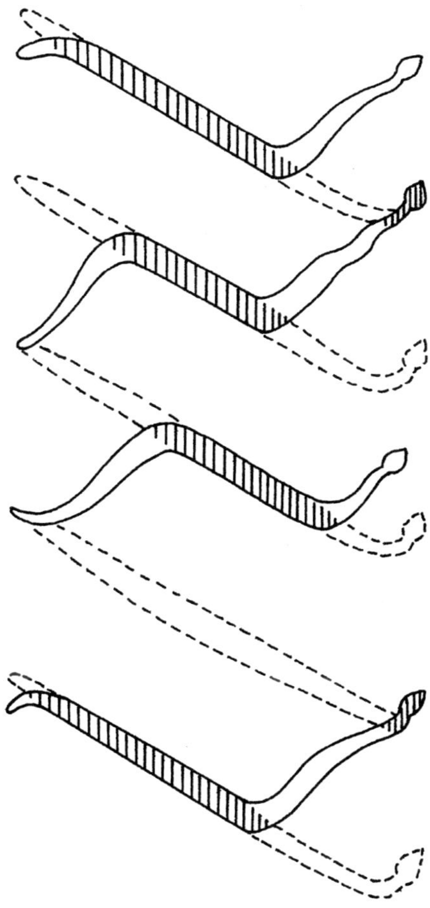

Sidewinding Technique

Housing

CAUTION!
Snakes are fantastic escape artists, and as such one must take every precaution to ensure that there are no escape routes.

This section is dedicated to housing snakes. There are several popular methods of housing a snake which are determined by the individual's taste or more correctly, by the requirements of the snake itself. Snakes have fairly basic needs when it comes to housing, the major influences in cage design are suitable size, ease of cleaning, maintenance, durability and cost. The majority of snakes are not particularly active creatures and approximately 0.5 metre^2 per 0.5 metre length of snake is a suitable size for the vivarium.

Ventilation

Irrespective of the type of vivaria selected to house your animals, each vivarium should allow for adequate ventilation to promote good health. Ventilation may be provided by cutting holes or slots in the sides or rear of the units, preferably at the bottom and top of the vivarium. With two ventilation holes the warm air rising pulls in fresh air through the bottom air vent and expels the waste air through the top. The ventilation holes should be covered with a fine wire mesh to prevent any escape attempts. Alternatively use the plastic ventilation grilles available from most D.I.Y. outlets. If using the professional grille, place a piece of fine wire mesh on the inside of the vivarium to prevent the snake within pushing the grille out of place.

Types of Vivaria

Glass vivarium (Type A)

This is probably the cheapest and easiest to clean vivarium that you can purchase. Since all the sides (excluding the cover) are made from glass, cleaning time is reduced. However, they have the disadvantage that heat is lost through the glass walls and a secure lid or top must be fixed. To overcome the problem of heat loss, glue or sellotape pieces of polystyrene to the outside of the tank, on three sides only. As for a lid, look at purchasing the various professional vivaria lids which have been designed specifically for glass aquaria and for the keeping of reptiles and amphibians. These lids are available from most pet stores.

A typical glass tank fitted with a vivarium lid.

The vivarium lid can be fixed to the top of the unit using clear silicone based glue available from most pet stores and D.I.Y. type shops. The lid provides a glass sliding top for maintenance work, a metal grill providing ventilation and a hole for a light bulb holder, although the authors would tend to feed a heater pad through this opening.

Wooden Vivarium (Type B)

This is perhaps the most common vivarium consisting of wooden sides and glass panels. These can provide access in the form of sliding doors, or possibly fixed glass as a means of viewing the interior of the vivarium. Since the main construction is plain wood or laminated chipboard, the vivarium can be fabricated to fit into any section of a room. Vivaria of this construction can also be inexpensive to make, and can be fabricated from household furniture. With constructions using laminated chipboard, all surfaces are easily cleaned using a damp cloth and suitable disinfectant.

A wooden vivarium with glass lock, light and vent.

The basic construction consists of wooden or chipboard panels screwed together to give the bottom, top and sides, a plastic runner for the glass (if glass doors are to be used) and the correctly cut glass with all sharp edges removed.

The vivarium should be made quite tall and narrow for arboreal (tree dwelling) species of snake, and wide and low for ground dwelling snakes. It is possible to place a false floor in the bottom of the vivaria through which is drilled a hole large enough for the snake to pass through. This provides the snake with a retreat if it so desires. Within this false floor is placed a drawer arrangement which allows the keeper access to the snake once it has retreated. This construction would also have an access from the front to allow the vivaria to be cleaned.

Glass doors in the front of the vivarium allow easier access for cleaning and feeding, others have fixed sides and a lid on the top for entry.

The wood or chipboard (if not laminated) should be treated with a good varnish to prevent the wood from becoming damp through the snake's faeces and water spillage. All corners and joints in the vivaria should be sealed with silicon sealant to prevent leakage. If using glass doors they should be fitted with a suitable lock and key arrangement to prevent the animals accidentally opening the vivaria and escaping.

Plastic Boxes (Type C)

Plastic boxes may be used to house both hatchling and adult specimens alike. As these boxes are available in many shapes and colours, they can be purchased to blend in with any other equipment you may already have.

If using plastic boxes they may be stored within a larger vivarium which is itself temperature controlled. Using this technique only one set of temperature control equipment need be purchased to allow several specimens of similar requirements to be maintained.

Alternatively the plastic boxes may be mounted on shelves with a simple heat strip or heating cable run along the rear of the shelf. With this method, the animals move to the rear to absorb heat and to the front to cool down, thus ensuring thermo-regulation. The heat strip or cable must be controlled via a suitable thermostat or via a 'dimmer-switch' which varies the amount of power applied to the cable, thus adjusting the amount of heat generated.

Plastic boxes are particularly useful for those species that are very secretive by nature or problematic feeders.

Plastic boxes are ideal for keeping hatchling snakes as in this display case at Coventry Reptiles.

Vivarium Furnishing

There are a few basic rules and guidelines for furnishing the vivarium.

> **CAUTION!**
> **'THINK OF THE SNAKE'S REQUIREMENT FIRST'.**

Substrates

All vivaria must have a substrate i.e something covering the bottom of the unit to soak up faeces and water. Snakes need to be kept on dry substrates otherwise they have a tendency to develop scale rot and other fungal diseases. Even Dice snakes and Water snakes, which enjoy a damp natural habitat, only require a water bowl with the remaining floor area kept dry.

Whichever substrate is used, it is essential that waste uric acid or faeces are removed as soon as possible, as these provide a growth medium for many strains of harmful bacteria. There are many substrates that you can choose from.

Paper: (Code A)

This is possibly the substrate most frequently used by the amateur and professional herpetologist alike. It has several advantages in that it is cheap, and in the case of newspaper there is some evidence to suggest that the chemicals used in the inks kill certain bacteria, and contribute to the well being of the animals. In addition the paper allows rapid spot cleaning and absorbs faeces and water spillage. To our knowledge there is no evidence to suggest that snakes kept on newspaper or plain paper have suffered any side effects.

The disadvantage of using paper is that it is not as attractive as other forms of substrate, and snakes tend to bury themselves between the sheets thus destroying the aesthetic qualities you may have created only minutes previously.

Sawdust: (Code B)
Some herpetologists use sawdust as a substrate. Sawdust has exceptional absorption qualities in soaking up water and faeces, however, it suffers the drawback that should it be swallowed by the snake, it may swell within the stomach leading, under worst circumstances, to death unless veterinary assistance is sought.

Bark-Chippings: (Code C)
Bark chippings look very attractive in the vivaria and produce a more realistic appearance, however the authors feel that bark chippings have more disadvantages than advantages to the herpetologist. Firstly freshly imported bark is often infested with mites which require the bark to be placed in a confined area with an insect-killer strip prior to its use. Secondly, it is inadequate in soaking up faeces and water spillage and camouflages any faeces. When faeces are spot cleaned it never removes all of the bacteria unless the complete vivaria is disinfected.

Gravel: (Code D)
Another popular substrate though very limited in its use. Gravel looks natural, but again its disadvantages out-weigh its advantages. It must be regularly washed to prevent bacterial build-up and it does not soak up faeces and spilt water. Should the snake accidentally swallow some gravel with its meal it can lead to serious problem which may require medical attention.

Wood shavings: (Code E)
Wood shavings are frequently used by herpetologists being readily available and relatively appealing to the eye. It has the advantage of its absorbency properties, but the disadvantage of wood preservatives, splinters and dust which may cause harm. If the shavings are swallowed they may cause an internal blockage. If using wood shavings, remove the dust and splinters through a sieving process before placing within the vivaria. It is now possible to purchase laboratory grade shavings which are preservative free.

Sand: (Code F)

Sand should only be used for those species naturally occurring in desert habitats. The best sand to use is coral sand obtainable from aquatic pet stores. NEVER use sand with species that do not naturally live in a sandy environment, as the sand tends to become lodged under their scales, causing irritation and skin infections. In addition if large quantities of sand are ingested with the food it may cause injury to the alimentary canal.

Peat: (Code G)

Used by many herpetologists (commonly with a little sand mixed in). Peat soaks up water and faeces, is quite easily cleaned and looks quite attractive particularly if a little moss is allowed to take root. However, peat has the disadvantage of holding moisture and snakes kept on peat over an extended period of time may develop forms of scale rot and other skin complaints.

Hard Wood Chips: (Code H)

Again quite attractive within the vivaria set-up. They have limited absorption qualities for soaking up water and faeces, but are available mite and insect free. Again they have the disadvantage that if swallowed it could result in medical attention being sought in order to rectify any problem.

Corn Granules: (Code I)

The late 80's saw a trend in herpetologists using either corn or oat granules as a substrate. On the positive side both are available in a range of sizes, they are both aesthetically pleasing giving the vivaria a more natural look, both absorb water and faeces and allow spot cleaning to take place. On the negative side, if the substrate level is relatively deep the snake's movements around the vivarium will sometimes cover water spillage and faeces which invariably lead to mould growth on the granules. If used, the substrate should be replaced in full at least every 3 - 4 months or at the first signs of mould growth.

Carpet Tiles: (Code J)

These can be more attractive in appearance than paper. They should be changed whenever soiled with waste products. One possible solution is to have several spare tiles which can be substituted for the soiled examples which are then washed and used as necessary.

Cat Litter: (Code K)

Cat litter has been used by several herpetologists with varying degrees of success. Available in several different grades, the larger grade is suitable as a substrate for most colubrids, although care must be taken to ensure that none of the litter is ingested with the snake's food. Cat litter can be attractive and has the advantage of soaking up faeces and spilt water, however any spoilt areas must be cleaned as soon as possible.

Internal Furnishings

Climbing Branch and Rock

Each vivarium should have a log, branch or rock which the snake can use to aid sloughing. If possible both a rock and branch should be included. Before placing these in the vivaria, ensure that they have been washed in warm water containing a bacterial disinfectant, such as Miltons, to ensure that no bacteria will be transmitted to the cage.

Hidebox

A hide box is a vital piece of furniture with certain species, particularly those that are secretive and have difficulty in adjusting to captivity. In their natural environment a snake would spend much of its time undercover. Replicating this effect with a hide box provides the snake with a form of security against predators.

The hide box must be large enough to permit access by the snake, and small enough so that the snake's body touches at least three sides. A hide box can be as simple as a cardboard box or more decorative like a piece of cork-bark. It can also be made from rocks and slabs of concrete.

It has been shown that clear perspex hide boxes also provide the animals with a sense of security probably due to body contact with the walls of the box.

Water Bowl

A water bowl is one of the most important items in the vivarium. Consider using a ceramic bowl, which are extremely useful as they are less likely be tipped over. Margarine tubs have been found to cause problems through spillage, and if lids are fixed to the tubs with a hole cut for access, snakes have been known to drown when they became trapped within the tubs, unable to escape.

Fresh water should be placed in the vivaria at least every other day or more frequently if the animal defectaes in the bowl.

This Californian Kingsnake is happy in its plastic box, with a water bowl and paper tissue as a substrate.

Artificial Plants

> **CAUTION!**
> **Do not to use real plants in the vivarium as they are often destroyed by the snake within a very short period of time.**

Artificial plants make the vivarium look a little more attractive. There are several different types of plastic plant available to suit the individuals taste and pocket, available from most garden centres and herpetological dealers.

Essential Equipment

There are a number of essential pieces of equipment necessary to ensure that the snake remains at its optimum temperature. In essence this simply consists of a thermostat, some form of heating element and a thermometer.

Thermometer

A very simple device purchased from pharmacists, electronics suppliers and pet stores, which allow the internal temperature within the vivarium to be monitored. If possible fix two thermometers, one either end of the vivarium to measure the thermal gradient across the vivarium.

Thermostats

The thermostat is set to switch the heating element on and off at a given pre-set temperature. There are several currently available to the herpetologist varying in price and complexity.

Bi-metallic Strip Thermostat:

Used by many amateur herpetologists as a cheap alternative to expensive electronic stats. They operate on the bimetallic strip principle, whereby two pieces of metal bend according to tho temperature, making and breaking the connection to the heating element. These thermostats tend to allow the temperature to fluctuate by several degrees around the set point and must be mounted within the vivarium. They also tend to be cheap and unreliable.

Electronic stats:

Probably the most common thermostat used, allowing a great deal of control over the temperature setting. The thermostat is mounted outside the vivarium and a temperature sensing probe is run into the vivarium to monitor the internal temperature. This type of thermostat switches the electrical supply to the heating element electronically.

It is slightly more expensive than the bi-metallic thermostat already described, however, the internal temperature of the vivarium is controlled to within 1°C of the pre-set temperature required.

Modern Electronic thermostats allow temperatures to be controlled to a high degree of accuracy

Sophisticated electronic stats: Several types of electronic thermostats exist which do not switch the heating element on and off, but rather vary the electrical power supplied to the heating element gradually. This action ensures that the heating element lasts longer and keeps the temperature very stable often within 0.1^{o}C of the pre-set temperature. However, it has the disadvantage of the relatively high cost. Some people also argue that the reptiles involved do not need such accuracy in temperature setting due to the fact that in their natural habitat the temperature would fluctuate constantly.

In addition, some of the more expensive electronic thermostats also offer a range of other specialist options. Several allow for different temperatures to be selected for the daytime and night, others allow several vivaria to be controlled from one unit, whilst some also allow not only the heating source to be regulated but also the lighting within the vivaria, allowing for the correct day and night lighting ratio to be maintained, for example 16 hours of daylight in the summer and 8 hours only in the winter.

Heating Equipment

As with other pieces of equipment, there are several methods of heating the vivarium according to the snake's requirements and your preference. An important consideration in selecting the correct heating element, is to allow the snake to thermoregulate by placing the heating device at one end of the vivarium only. This allows the animal to move into and away from the heat source as required.

The incandescent lightbulb: (Code A)

The incandescent lightbulb: A popular choice of heat amongst the novice keepers.

The common lightbulb is used by many to provide heat, as they waste so much electrical energy in creating heat rather than light. It has the advantage in being readily available and inexpensive to purchase, however there are several disadvantages. The main disadvantage is that light bulbs tend to 'blow' frequently, leaving the reptile without heat unless detected quickly. Also whilst the light bulb is switched or dimmed some herpetologists believe this could cause stress to the snake although no scientific evidence has shown this statement to be correct.

Snake

If the decision is made to use light bulbs we would suggest one uses a red or blue bulb which is less likely to cause stress during night-time. Rather than using normal light bulbs, one should consider the use of the various ranges of spot lamps available, which will tend to provide a more definite hot spot beneath the spotlight. The lightbulb should also be protected using wire gauze, which not only prevents the snake burning itself on the bulb, but also prevents the bulb being dislodged.

Ceramic Heaters: (Code B)

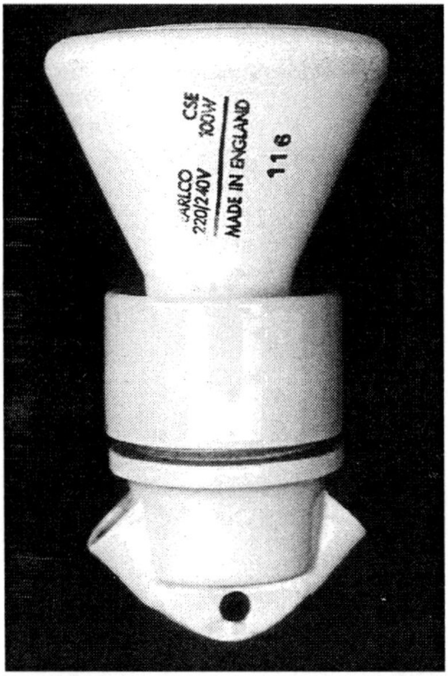

The Ceramic Heater.

Available as 'trough' or conical (resembling a lightbulb) in shape. Ceramic heaters must be mounted with special brackets, and a wire gauze safety cage fitted over the heater to prevent the snake burning itself. Available in a range of wattage sizes suitable for the larger vivaria, these heaters are very robust and will last for several years.

As with other forms of heating elements there are some disadvantages; firstly, if a low quality thermostat is used and a fault occurs resulting in the element being constantly switched on, the ceramic heater can overheat the vivarium with the resultant loss of the snake and a possible fire. Secondly, so much heat is dissipated from these types of element that in vivaria smaller than 1 metre in length, the reptiles within are not provided with a thermal gradient, which may cause animals to stop feeding.

Heating Cables: (Code C)
Available mainly from garden centres in various lengths, heating cables are ideal for heating several banks of cages in a stack unit or for heating 'shelf-type' displays. Due to their low wattages, they are not so dangerous should the thermostat fail, and allow for thermoregulation if only the rear section of a vivaria is heated by the cable. Heating cables tend to be slightly more expensive as one must consider the added cost of safe securing clips or tape. In larger vivaria, the cable should be mounted internally and clipped securely in place.

Heat Pads: (Code D)

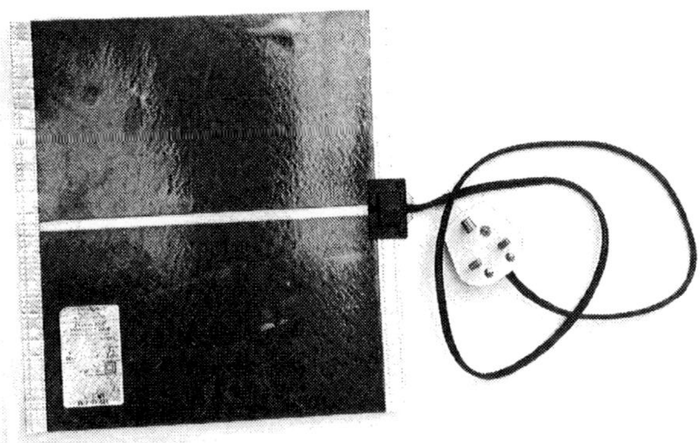

The Heating Pad.

Available from most pet stores as 'Ultra-Therm' or 'Micro-Climate' heat pads. A range of sizes exist both in physical dimensions and wattage rating. These are ideal heaters providing localised 'hot-spots' within the vivarium, upon which the snake may lay to digest its food thus providing the reptile with a choice of substrate temperature from which it can thermoregulate. Due to the fact that the wattage sizes available are relatively small (often 10 - 40 watts), there is a reduced risk of overheating should the thermostat fail.

A recent addition to the range are a family of long slim pads up to 2 metres in length (longer by special order) which act in much the same way as heater cables and are particularly useful for heating rows of plastic vivaria.

Micro Ceramic Heaters: (Code E)
These are the latest in heating equipment at the time of writing, combining several technologies. The units resemble ceramic heaters but have the advantage in only providing infra-red heat at relatively low wattages, thus should a thermostat fail the vivarium is unlikely to overheat.

Available in several sizes and power ratings, the smaller units are ideal for the heating of vivaria for the colubrids. Another advantage is that the units are assembled with several elements, thus even though one element may 'blow', the remaining elements will continue to function. This type of heating element does not normally require a protective wire guard as they are supplied fully fitted within a protective box with air vents.

Tubular Heaters: (Code F)
Tubular heaters are available from the larger retail outlets and D.I.Y. stores often seen as 'frost protection' heaters for garages. Available in the sizes 1, 2, 3 and 4 feet, they provide a cheap alternative to ceramic heaters and are especially useful for heating vivaria housing large boids.

Hot Rocks: (Code G)
A novelty heater purchased from pet stores, these units are prefabricated rock look-alikes with a heating element built in. These units can replace the heat pad to give a localised hot spot.

Lighting

In recent years the type of vivarium lighting has taken on a greater significance after research in the U.S.A. has shown that light affects the reproductive cycles in many snake species. Several herpetologists have noticed that biological physiology or behaviour of reptiles can be influenced by the type and proximity of light sources.

Although eyes are not particularly well developed in snakes, the sense of sight brings the animal a large proportion of its information. The rising of the sun gives the light that shows shape, detail and colour to the world, while the changing of photo periods stimulate breeding activity and other hormonal effects which are brought about by light waves acting on the animal.

Light is a form of energy, but the light the animals sees makes up a small proportion of a much larger group of waves with different wave lengths known as the electromagnetic spectrum. As snake keepers we should not only be concerned with visible light, but also wave lengths of shorter values than those of light seen by the human eye.

Many experienced herpetologists believe ultra violet wave lengths are an important aspect of vivarium lighting, these wave lengths are radiated from very hot objects and are present in solar radiation from the sun. Many reptiles, lizards in particular, seem to be very responsive to effects of light containing waves from the ultra violet region of the electromagnetic spectrum.

In most vivarium systems two main types of lighting are in common use, these are incandescent bulbs or fluorescent tubes. Incandescent bulbs are cheap to purchase and available in many configurations and power ratings. They are not a particularly efficient form of lighting as they also produce a large heat output. They may, however, be a useful means of warming the vivaria as well as lighting it as previously explained.

Many snake keepers with large vivaria are now using full spectrum fluorescent tubes. These produce wavelengths of light which are very similar to the natural rays from the sun, the tubes produce not only visible light but also electromagnetic wave lengths in the ultra violet region.

Ultra violet emitting tubes affect calcium uptake from the diet of animals, if calcium is not absorbed from the animals food it is removed from the skeletal bones and the bone collapses. U.V. waves bring about the formation of vitamin D3 which acts upon the reptile's gut and causes the uptake of calcium from the diet.

The fluorescent tubes currently marketed as simulating natural sunlight rapidly lose their efficiency at producing U.V. waves and so need replacing at regular intervals, at least every 12 months. They also require bulky control/starter units which are an additional cost.

Many snake keepers have, however, successfully kept and bred snakes without access to U.V. producing light sources. Several species suitable for beginners are crepuscular or nocturnal and are rarely found basking in bright sunshine in their natural habitat.

If the species of snake which you intend to purchase is diurnal then it may be beneficial to consider full spectrum lighting. Recent research has indicated that several species of reptile feed more readily if exposed to natural lighting systems.

Manipulation of photo periods, the ratio of light to dark, has been shown to be an important factor in the synchronisation of breeding in many species of snake. In particular snakes from temperate regions are brought into breeding condition during shortened daylight hours, as spring approaches and daylight hours lengthen the males and females are stimulated into mating.

In captivity the photo period can be artificially manipulated by the use of time switches attached to the vivarium lighting system. During summer months a suitable ratio is 14hrs light:10hrs dark, during winter this can be reduced to 8hrs light:16hrs dark. If this procedure is used in conjunction with temperature changes many species which have proved difficult to breed in the past may be persuaded to reproduce in captivity.

Purchasing and Quarantine

Purchasing a snake

Most individuals, if purchasing a snake for the first time, tend to approach a pet store. We recommend that you should make a point of looking around, asking the advice of other experienced herpetologists and only deal with those trades people that have a good reputation. Where possible buy only from those outlets that specialise in reptiles, as they provide more accurate information and have all the necessary and associated equipment that you will require.

An unfortunate aspect of the not so desirable pet stores is that some pack animals together in inadequate tanks which may result in mass spread of infection, mites and ticks.

External injuries can normally be treated and heal fairly quickly but often leave ugly scars. Never purchase an animal with an 'oozing' wound. Check the mouth for signs of 'bubbling', excessive amounts of saliva, mouth rot (rotten looking gums often white or yellowish in colour and occasionally with signs of blood) and injuries caused by rubbing its snout against the vivaria. In the case of mouth rot, even experienced herpetologists can have problems in successful treatment.

Never purchase an animal that appears limp and non responsive. Ignore comments such as the snake is resting and becomes more lively at a later hour.

THE GOLDEN RULE:
IF IN DOUBT DO NOT PURCHASE

If you are not sure ask someone you trust that has some knowledge on the subject or contact one of the societies listed in the appendix. Most importantly again, only purchase from a shop which is used by experienced herpetologists.

Where possible, only purchase 'captive bred' specimens. Captive bred animals are generally more expensive than those imported, but usually lack the diseases brought in by imported animals, and tend to have a higher survival rate. Most experienced herpetologists tend to purchase only captive bred stock from reliable breeders (often someone they have known for several years).

We suggest that you purchase a young snake, either a hatchling or a yearling, as these animals are more likely to live longer if maintained correctly. This gives both of you a chance to get used to each other right from the start and in addition you will enjoy watching it grow into an adult and take pride in your success.

Transporting your Snake

Transportation of animals is a subject very often overlooked in literature, yet this small detail can account for the loss of your snake even before you have had the opportunity to place it within your set-up at home.

For short distances, directly from the place of purchase to home, you may safely place the snake within a suitable bag (pillow-case for colubrids and large bags for boids). The bag should be placed on a flat surface, possibly on your lap. Avoid direct sunlight and temperature extremes, for example warm boots of cars. The bag should be prevented from sliding around to avoid damage and stress to the animal. Once home place the bag in the quarantine vivaria and leave the top of the bag open. Either allow the snake to come out itself or very gently tip the snake out into the vivaria, being careful not to allow the snake to shoot out of the vivaria over the top of your hands.

For longer distances place the animal in a suitable bag as described above. This bag is then best placed in a wooden vivaria packed half full of polystyrene beads and where possible use a heating bottle or warming element. Take care that the temperature is not excessive and is tolerable to the animal. Once the bag has been placed onto the beads, a thin layer of beads should be placed on top of the bag and the lid of the container secured tightly. Obviously adequate ventilation is essential to ensure the well-being of the snake. In addition clear warning labels should be fixed to the outside of the container detailing the contents and any other special messages e.g. 'Keep this way up'; 'Store in a warm, dry area'; 'Live animals'. Your telephone number both at home and at work should also be detailed in case of emergencies.

At the time of writing, a recommended courier for the transportation of reptiles is AMTRAK.

Quarantine Guidelines

For those of you that have kept reptiles for several years, the fundamentals of quarantine procedures for newly acquired animals should be familiar. However for those that may not be up to date or for those with little or no experience in herpetology, we shall now discuss the basic guidelines to prevent, as far as possible, the introduction of parasites and disease to existing specimens within one's collection. A hobbyist just about to purchase their first animal obviously has no quarantine problem if only one animal is to be purchased.

When a new animal is added to an existing collection (even if only one other animal), it should first be quarantined for at least three months or longer (the longer the better). All new charges should also be totally isolated, kept in a separate room, or moved to the opposite side of the room.

Quarantine vivaria should be sterile and contain only the bare essentials. The essentials are of an easily removed substrate (paper), a climbing branch that has been stripped of bark to prevent any parasite being carried into the vivarium, a rough surfaced rock to aid sloughing and a drinking bowl.

Optimum temperature conditions should be maintained. Regular, thorough disinfection of all accessories, handling and feeding equipment is also very important. At this point it should be noted that ordinary household pine disinfectants should be avoided as they can irritate or kill the animals and also fail to kill many of the fungal infestations. Use those disinfectants supplied through your local veterinary practitioner, pet store or pharmacy, for example Miltons or other sodium hypo chlorite based solutions.

Prior to a new purchase being placed in the quarantine vivarium it should be bathed in warm water to remove dirt and any other particles attached to the scales. One should ensure at this point that the cloaca is free of hardened pieces of faeces and solid uric acid particles. Whilst the animal is being handled, it should be thoroughly inspected for any injuries and for ectoparasites (mites and ticks).

When the animal is placed in the quarantine tank, the faeces play an important part in helping to determine if any internal complaints are present. If the faeces are of a mushy consistency and of greyish coloration, or if the animal refuses food or regurgitates partially digested, rather smelly odorous prey animals after a few days, this can indicate an acute gastro-enteritis or infection of the digestive tract. We would suggest that such faeces be collected and taken to your local veterinary surgery for analysis as soon as possible.

The faeces of animals feeding on small mammals will be sausage-like, firm and a red-brown to black in colouration. If newly hatched chicks are used, the faeces will be rather mushy and smell quite badly, however, the food should never be regurgitated. With fish, the faeces is rather like that of chicks, being mushy and smelly. When an animal produces waste, the faeces and uric acid should be removed immediately as some parasites rely on eggs being transported to a second host via the faeces. The tank should be cleaned at the same time.

If ticks are observed on the animal, the tick should be covered in Vaseline which will deprive it of oxygen. When the tick has asphyxiated (leave under vaseline for 24 hours), it should be removed with a pair of tweezers. The tick should be grasped about its mid-body and with a side to side movement dislodged from the animal ensuring the mouthparts are also taken. *Never* pull the tick straight from the body of the snake as this often leaves the mouthpart of the tick embedded in the body which leads to infection. The area should then be cleansed with an iodine based solution, for example Betadine, to prevent any secondary infection occurring.

In the case of ticks (and mites), 'Vapona' or a similar insect killer strip can be used. This approach should be applied with care as damage can be caused to the snake if it comes into contact with the strip or if too large a quantity is administered. This method kills adult mites but does not affect the eggs that are deposited around the vivaria. The result is that mites suddenly reappear many weeks after it was thought they had been eradicated.

The insect strip should be placed within the vivaria for 3 days, housed within a small tin or cardboard container with ventilation holes, it is then be removed and the vivaria cleaned. This process should be repeated three times with a one week interval between each insertion of the insect strip. This technique should eliminate young mites that hatch between insertions. During the time that the Vapona strip is in the vivarium, the water bowl should be removed otherwise the Vapona particles will dissolve into the water causing potentially lethal results should the snake drink excessively.

A series of reports published recently suggest that Vapona can actually result in the death of the snake within the vivaria if used excessively i.e. left in the vivaria for several months. Thus *never* leave Vapona in the tank for longer than a three day period, at any one time.

Another method without the use of insect killing strips, is to change the substrate and wash the walls of the vivarium and accessories every 2 - 3 days. The animal may also be bathed. If perforated cardboard is placed inside the vivarium to act as a hide-box, ticks tend to congregate in the corners, perforations and folds of the cardboard to lay their eggs. With the cardboard taken out every two days, the mites and their eggs are removed together.

Finally at the end of the quarantine period, after the animal has had a thorough examination, it may join the rest of the collection. All accessories within the tank, excluding the water dish and temperature related devices, should be discarded, and the tank disinfected ready for the next animal.

Do not use household pine disinfectants as these are poisonous to reptiles.

Dietary Requirements

There are many items available from your local pet stores that may be benificial for, or detrimental to your snake. It is easy to get confused about the correct dietary requirement of your animal, with the result that the animal's health deteriorates over time. We shall describe the various food items sold, giving our opinion as to their suitability. However, it should be stressed that the information presented must only be used as a guide and one should continually investigate the possibility of other sources of nutrition.

Rodents

This category is by far the easiest to obtain and more often the best part of a staple diet. Rodents, comprising mainly mice and rats, are rich in protein and would normally constitute the snake's main prey item in captivity. Rodents are freely available frozen through the pet-trade, however always ensure your food animals are of a high quality and disease free.

Many experienced herpetologists believe that if your snake is normally fed mice and rats, then a gerbil fed every 20th meal will help clean the intestines of the snake due to the much coarser hair. This is evident by the fact that much more hair is found in the faeces than one would normally find with faeces passed after offering mice and rats.

Fish

Fish are an essential part of any aquatic or semi-aquatic species (Garter snakes and Water snakes). Goldfish and other fresh-water fish can be purchased from pet stores, or 'Whitebait' and other similar fish can be used provided that they have been thoroughly washed and par boiled for 30 seconds to destroy enzymes that would otherwise cause a Thiamine deficiency. Vitamin supplements should be used with all meals to ensure a healthy balance of vitamins and minerals.

Birds

Hatchling birds are other common items captured by snakes in their natural habitat. Avoid using chicks as there is little protein and calcium for the snake. A snake raised entirely on chicks is more often out of condition and considerable work is necessary to raise the animal to a healthy breeding size. Adult birds offer much needed protein and nutrients. The size of adult birds can range from quails, to ducks and chickens.

Amphibians

There are many snakes, particularly those that normally take fish, that relish the odd frog. There are stories of grass-snakes and other semi-aquatic snakes refusing all food until presented with a frog. One of the authors has experienced an Eastern Milksnake, Lampropeltis triangulum triangulum, refusing all food until a frog was rubbed over a mouse pink. Frogs are not readily available from shops, but should not be collected from our natural habitats. It is recommended that you raise a group of frogs yourself and simply rub the prey item along the back of the frog (much to their annoyance) before offering the prey to the snake.

Reptiles

Kingsnakes and milksnakes are ophiophagous i.e. they will take other snakes as part of their natural diet. Always be aware of this important piece of information to prevent your cage-mates eating one another. Some snakes only eat specific species of lizard in their natural environment and again this can be overcome through the use of 'odour-manipulation' which is dealt with in greater detail in the following chapter.

Garter Grub

The '90s saw the introduction of specialised food, especially formulated to provide the correct balance of vitamins, calcium and nutrients for Gartersnakes and their close allies the water snakes. Sold under the trade name of 'Garter Grub', the meals are provided in fish-finger type shapes which are defrosted and offered directly to the snake. This form of food package does not require additional supplements to be used and provides the keeper with an easy to use food.

Supplements

Provided that a balanced diet is offered then your animals may never require any other vitamin or calcium supplements. However, as a general rule it is suggested that a small amount of vitamin and calcium additives be placed in the mouth of the prey item at least one in every six meals.

Feeding

This section is dedicated to feeding and methods of persuading non feeding snakes to resume a healthy feeding program should they have any difficulty. One thing you should note if you have never kept a snake previously, is that a snake is not fed daily but rather weekly or monthly as they are poikilothermic.

Preparation of food items

Unless there are exceptional circumstances, every meal you obtain for your snake will be frozen. Before the prey item is given to your snake it must obviously be defrosted and brought up to at least room temperature.

The best ways of doing this are by soaking the prey item in a bowl of luke warm water for several hours, or to simply leave it to defrost in a warm room for at least 12hrs - 36hrs (depending on the size of meal to be offered). The item should be checked to ensure that no frozen matter remains, particularly in the gut section, and that the animal is not cold.

Under no circumstances should prey items be defrosted in the microwave or by placing them directly onto a heating element such as a radiator. Items defrosted this way are generally refused and tend to smell quite badly, whereas items defrosted as instructed will have a more natural odour.

Normal Feeding Routines

When your animals are feeding regularly, the following guidelines should be used to ensure their successful acceptance of food and a long life.

- Never handle the food if it can be avoided. If you have to handle the food it should be washed in warm water before being placed in the vivarium.
- Only offer food using forceps, not by hand.

- Never feed two animals of dissimilar size at the same time in the same vivarium. If you are ever in the situation of having two animals with a hold on one prey item, place them both in a bucket of luke warm water to separate them. Alternatively squirt a small quantity of detergent or alcohol into their mouths. This does no harm, but should make them release immediately.
- Never leave live food in a vivarium unattended and certainly not over-night. You should avoid using live food at all costs.
- Never over-feed your animals. This is perhaps the most common mistake amongst herpetologists (even experienced herpetologists). One should note that once fat has built up, it very difficult to shed and the animals are less likely to breed. Hatchlings (kingsnakes, milksnakes, ratsnakes etc) should be allowed 3-5 years to grow to sexual maturity. Hatchlings should be fed once a week to once every two weeks, depending on the size of food offered.
- Use supplements where necessary to ensure the snake receives enough vitamins. The vitamin supplements suggested by leading veterinary surgeons and herpetologists in the United Kingdom are Vionate or Vetark. These are readily available from veterinary surgeons, pet stores and some pharmacists.

Problem Feeders

Like all herpetologists we have encountered animals which have refused to feed. Snakes in general refuse to feed for a number of reasons, for example:

- Unsuitable environment.
- No cover or insufficient security.
- Human scent on the prey item.
- Unsuitable prey presented.
- Known problematic species.
- Specimen wild caught, adjusting to captivity.

Any one factor or a combination of these factors may be responsible for the problem. Over the years, experienced herpetologists have developed and handed down a number of techniques to encourage regular feeding patterns.

Environment

When a snake refuses to feed, particularly if newly acquired, first check the environment within which the animal is placed. The snake should be maintained at the optimum temperature for the respective time of year, have access to fresh water and be provided with a 'hide box' to offer security. In addition the vivarium should be placed within a suitable location away from the disturbance of children and other animals.

With hatchling snakes, one technique involves isolating the animal in a sandwich or plastic box with ventilation holes cut in the roof. A small hide box and water dish is placed in the container. Using this technique, many hatchling snakes can be induced to eat by the added security.

Food Item

Each snake should initially be given its normally prescribed food in captivity. Boids (boas and pythons) should be given mice, rats and rabbits, whereas Water snakes should be given frogs and fish. Colubrids (milksnakes, kingsnakes, ratsnakes) are best started on mice and small rats. If the normal food is refused, the following should be tried:

- If using mice and rats, try different coloured animals as some snakes prefer a particular colour prey.
- Try using fresh killed animals and animals that have been frozen, again some snakes have a preference for one or the other. One of the authors has two African pythons, Python sebae, in the same vivarium; one only eats fresh killed prey, whilst the other prefers a rat that has been defrosted and left to stand for at least 24hrs.
- Try other food items such as fish, gerbils, guinea-pigs, lizards and even other snakes as a last resort. A Reticulated python, Python reticulatus, only resumed feeding after 12 months when offered a dead king snake!
- If using mice and rats, try cutting the end of the nose

and/or the whiskers off the dead animal before placing it in the vivarium.

- If feeding kingsnakes and milksnakes which feed on lizards and snakes in their natural environment, rub the mouse or 'pinky' against a lizard before placing it in the vivarium. Alternatively, if dead lizards are available cut the lizard's stomach area open and rub the prey into the lizard's internal organs before feeding. The snake then picks up the scent of the animal in which the prey was rubbed or dipped and will often proceed to eat. If this process is repeated the animal is slowly converted to mice. This is often referred to as 'odour manipulation'. This method can also work with a pinky bound in the sloughed skin of a lizard or snake, or if the prey item is dipped into the egg of a reptile after the youngster has emerged.
- Before feeding a dead animal to a snake, soak it in warm water to improve the dead animal's smell.
- If using pinkies, try opening the skull to expose the brain of the dead animal. This can also add to the prey item's smell and attract the attention of the snake.
- Try using mouse tails and legs for hatchling snakes if pinkies are in short supply or are refused. A word of caution: A snake cannot receive enough vitamins from tails alone, hence the tail should be coated with the necessary vitamin and calcium supplements.
- Try to feed the snake at different times of the day and leave it alone for a few hours with the dead prey item before checking on its progress (this does not include live food).
- Try moving the prey item in front of the snake using blunt tweezers or forceps.

Hibernation

Wild caught north American specimens captured late in the year, will often refuse food and seek a secure hibernation spot as part of their natural cycle. Therefore, one should place them into hibernation to stimulate a feeding response. This technique is also very successful with new hatchlings that would normally think of hibernation if born in the wild.

Try placing the animals into hibernation by lowering the temperature to their normal den temperatures of 7°C - 13°C, for two to three months. One often finds they will then proceed to feed readily when brought back out of hibernation.

CAUTION!
Strictly monitor the snake's progress for loss of weight and condition during this period.

Force Feeding

In spite of all encouragement, a snake may still refuse to feed and start to lose condition. At this stage force feeding should be considered, but only after all other techniques have failed. We would like to stress that force feeding should only be attempted by the experienced keeper. If attempting to force feed for the first time, do so under supervision. Force feeding does not present a permanent solution, as many animals will commence normal feeding routines after an initial force feeding stage. There are a number of force feeding techniques available to the herpetologist:

The Pinky Pump

The pinky pump was developed in the U.S.A. and is basically a stainless steel syringe with a special tip fabricated to allow easier manipulation or insertion into a snake's mouth and throat. Pinkies are liquidised in a food blender with vitamin supplements and additional calcium where necessary, until a mushy 'pinky' material is formed. Alternatively one may use the many varieties of 'baby food' which are high in nutrients.

The liquidised cocktail is drawn into the pinky pump and the snake's mouth forced open by the end of the syringe. The cocktail is then gently squeezed into the throat using a very gentle but steady pressure on the syringe. Great care must be taken not to introduce a large a quantity of food too quickly in order to avoid air being pushed into the snake's throat. With the food in the oesophogous, the snake should be removed from the pump and the food massaged into the stomach.

This method can be used up to twice a week particularly with large animals, however, you must take into account the stress factor under which the animal is placed. If a pinky pump is not available a simple syringe and small piece of cathetor tube may be used to the same effect.

Other

Other methods of force feeding require the snake to be held whilst food is gently pushed into the snake's throat. When certain snakes are handled, their mouths automatically open (often in an attempt to bite) allowing food to be pushed in. If the animal does not open its mouth, then two techniques may be employed to achieve this.

The first is to try and open the mouth by very gently pulling on the lower jaw skin. The second technique involves the use of a slight bilateral pressure on the posterior jaw angles, or by a slight pressure between the lips using a soft wooden stick or rubber wedge.

With the mouth open, the food is placed in the mouth and gently pushed to the throat area. In some cases, the food being pushed between the teeth triggers a feeding response, however, it may be necessary to move the prey further into the pharynx. This will usually result in a swallowing action.

CAUTION!

If the animal would normally take a large mouse, then it should be fed one or two pinkies at one time when force fed. This method allows the snake's stomach to readjust to digesting the food, and increases the success rate in swallowing the prey.

Sex Determination

The ease with which snakes can be sexed by the herpetologist varies greatly from species to species, and individual to individual. Some find it very easy to sex a snake after a relatively short period of time, whereas other people who have kept snakes for several years still have great difficulty in determining the sex correctly.

Some of the techniques used to determine the sex of snakes must under no circumstances be attempted by the inexperienced keeper, unless accompanied by an expert in the particular method. Incorrect technique can lead to permanent damage to the animal involved, and may result in its inability to breed in future years.

There are several main methods employed in determining the sex of a snake which are deemed to be reliable. All of these techniques shall be described as all are useful within their own right. One thing that you should be familiar with before commencing, is that a male snake has two penis, each called a 'hemipenis'. The male hemipenis lies inverted pointing towards the tail (as shown in the illustration) and is exposed only for mating. The male will tend to favour one penis to the other. However, should one penis become damaged, the snake can function quite normally with the other.

Popping

This method is very useful for determining the sex of newly hatched or very young snakes up to 3 months of age. It should only be attempted by the experienced keeper. The method relies on the fact that male snakes have little control over the muscular action of the hemipenes until they are a few months old. With the tail slightly bent and a slow but gentle pressure applied from the tail tip towards the vent, the hemipenes of young males may be inverted i.e. pop out of the vent. This technique cannot be performed on adults or adolescents as they are strong enough to have control of the muscular region employed in inversion.

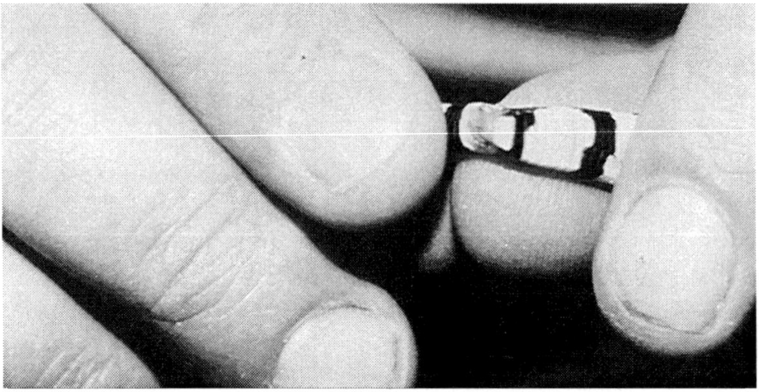

Popping a hatchling snake. Photo shows a female with glands exposed.

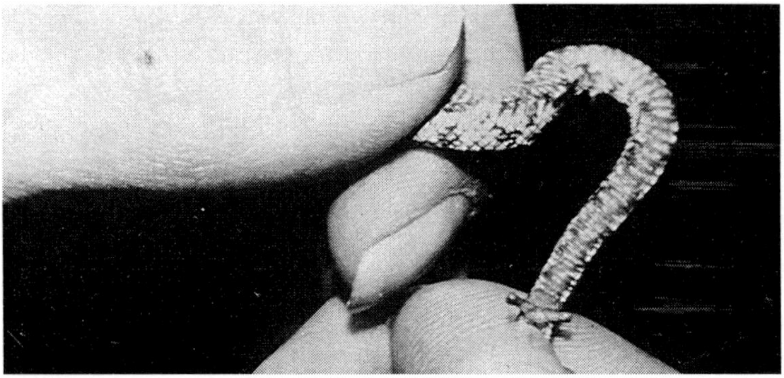

Popping a hatchling snake. Photo shows a male with hemipenes exposed.

In the hands of an experienced 'popper', this technique can result in 100% accuracy. However, it should only be used as a guideline until the animal's sex can be determined by one of the other methods which are deemed more accurate.

Probing

This is another technique of sexing which should only be attempted by the experienced herpetologist.

Since the hemipenes of a male snake are inverted within the tail, it is possible to insert a probe inside the inverted hemipenes for a distance several times greater than possible with a female. The distance varies from species to species, but in general it is possible to insert the probe several subcaudal scale lengths in the male (roughly 5 - 10), but rarely more than two or three in the female.

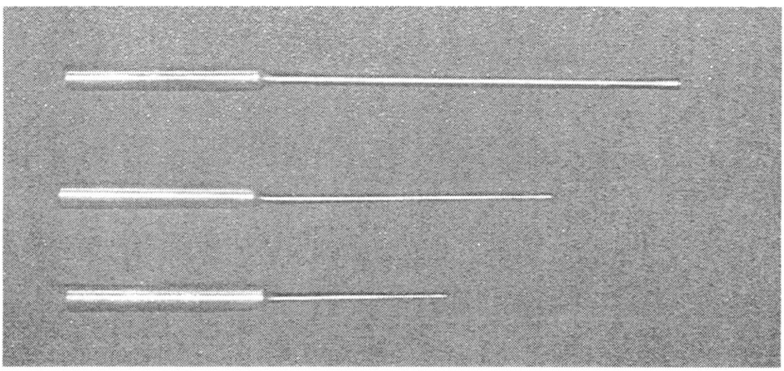

A set of stainless steel sexing probes.

The ball of the probe must be lubricated before use with Vaseline or a similar lubricant, and it is inserted gently with the greatest of care. Do not use lubricants unless clearly marked as non-spermicidal, as scientific evidence suggests that spermicides such as K.Y. jelly kills the male's sperm or reduces the fertility rate. If a probe does not slide in easily, then a smaller probe should be used. One must remember never to push the probe against any resistance, and be gentle so as to cause as little discomfort as possible.

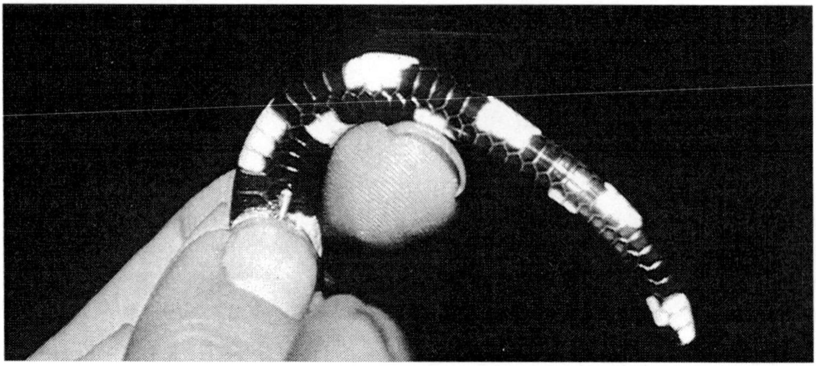

Probing a Californian Kingsnake, L.g.californiae. Shown is the depth at which the probe would penetrate a female.

CAUTION!
Inexpert probing can result in severe stress and physical damage to the snake which is likely to last for life. In the case of males, incorrect technique can lead to damaged hemipenes which may result in the animal's inability to mate. With females it can result in their secretion glands becoming damaged which prevents them releasing the necessary chemical stimuli used to attract and arouse the male.

Tail Shape

Some herpetologists are able, with experience, to determine the sex of a snake with reference to the tail shape since there are subtle differences between the tail shapes of the male and female.

In general the tail of a male is proportionately longer than that of the female and tapers with a shallow gradient, such that when viewed from the side the tail remains thick for some distance beyond the vent. There is also an enlarged bulge behind the vent that is far more pronounced in the male than that of the female. The bulge in the males is caused by the hemipenes.

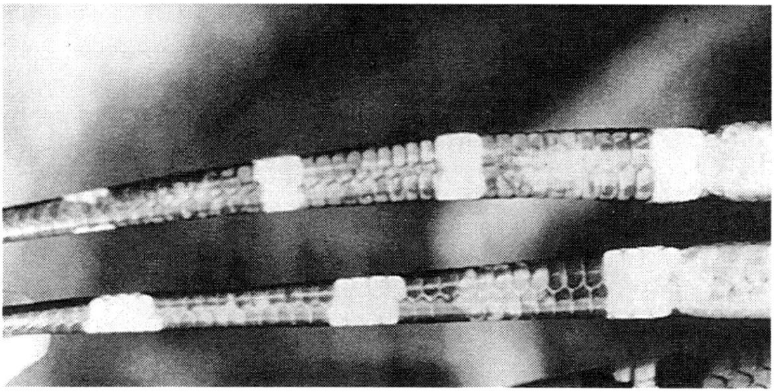

Sex determination through the visual inspection of tail shape in the Sinaloan Milksnake, L.t.sinaloae. Note the larger girth of the male in comparison to the female.

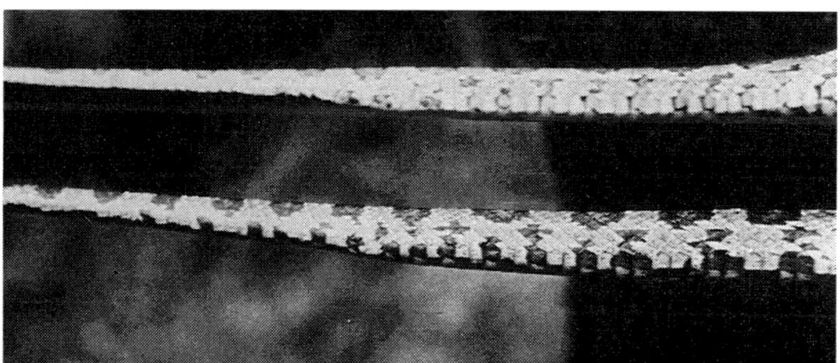

Side view of male and female Corn snake, E.g.guttata.

With a female viewed from the side, the tail tapers rapidly, resulting in a short stubby tail. When viewed from underneath, young females tend to have an evenly tapering tail whilst older specimens may have a bulge directly behind the vent. This is the female scent gland, which is used to attract the attention of the males over great distances in their natural habitat.

Snake

By running one's thumb and forefinger along either side of the vent and past the swellings, one can determine the sex (although not 100% reliably) by determining if the swelling resembles hemipenes or not.

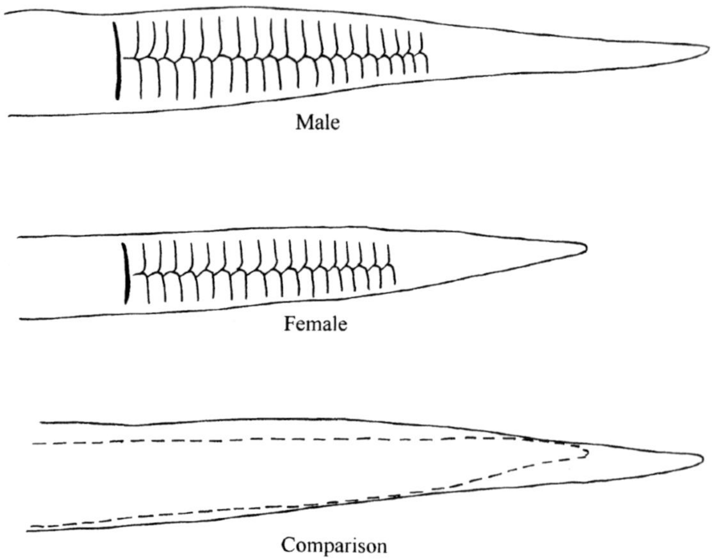

Illustration to show difference in tail shape.

Subcaudal Count

In most species of snake, the males have a proportionately longer tail than the female. Due to this phenomena males normally possess more subcaudals than the females. The subcaudals are the scales located from the vent to the tip of the tail on the underside of the snake.

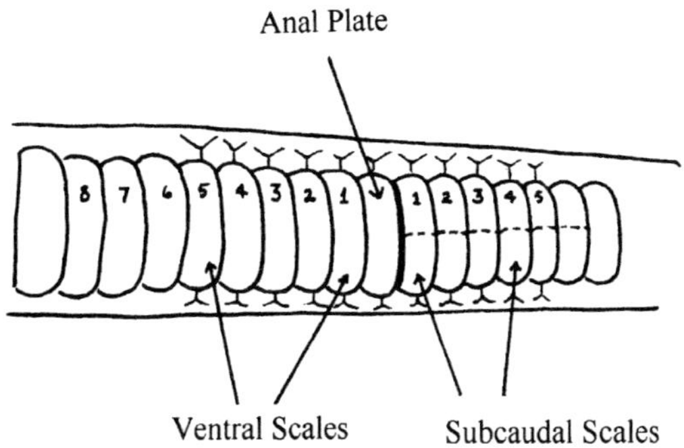

Location and technique for counting the subcaudal scales.

Although the number of subcaudals is not constant, it falls within a certain range for a given species. For instance, if the subcaudal range for a particular species is 40-90, the males are likely to have a subcaudal count of 65-90, whilst the female will lie in the range of 40-65. A subcaudal count in the middle of the overall range therefore, gives no real clue to the sex of the snake.

The subcaudal count method has the advantage for the novice herpetologist in that it does not require any previous experience and, furthermore, this method of sexing can be performed using sloughed skins, thus avoiding stress to the animal.

When referring to the list of subcaudal ranges, it must be remembered that they are the normal range and that counts outside those recorded are possible.

A list of common subcaudal count ranges are provided at the rear of this section.

Surgical Sexing

Surgical sexing or hormone investigation of faecal or blood samples is the most reliable method of sexing, and can be carried out by certain zoological gardens and research laboratories, although very uncommon in the U.K. This technique normally lies beyond the reach of the novice keeper. However, they should not be discounted if the means and facilities are available within the area. In the first instance consult your local veterinary surgeon for the samples to be taken and sent away for analysis.

Colour

There are a few species of snake that have a distinct difference in colour between the males and females, particularly in the vipers. All of the species described as suitable for the novice keeper in this book are similar when comparing the male with female's colour.

Spurs

With some species of snake, namely the Boids, the males often have longer and sharper 'spurs' just in front of the vent. The spurs are simply the residual hind limbs extending from the vertebrae. As a guide, if the spurs stick out approximately 1-2 scale widths and are sharp the snake is more likely to be a male. Females in general tend to have very short blunt spurs, often hidden beneath the scales.

Subcaudal count ranges for commonly available snakes

Scientific name	Common Name	Subcaudal Range
Boa constrictor	Boa Constrictor	49-60
Enydris cookii	Cook's Tree Boa	101-118
Eryx jaculus	Sand Boa	15-33
Python molorus bivittatus	Burmese Python	58-83
Python regius	Royal Python	28-47
Python reticulatus	Reticulated Python	75-102
Python sebae	African Rock Python	55-82
Lamprophis fuliginosus	House Snake	50-69
Coluber c. constrictor	Black Racer	72-107
Elaphe g. guttata	Corn Snake	48-79
Elaphe o. obseleta	Black Rat Snake	70-98
Elaphe o. quadrivittata	Yellow Rat Snake	85-100
Elaphe o. spiloides	Grey Rat Snake	70-89
Heterodon nasicus	Western Hognose	27-47
Lampropeltis g. getulus	Eastern King Snake	38-58
Lampropeltis g. californiae	Californian King	41-62
Lampropeltis g. floridana	Florida King Snake	40-60
Lampropeltis g. holbrooki	Speckled King Snake	38-55
Natrix tesselata	Dice Snake	48-79
Nerodia sipedon fasciata	Banded Water Snake	53-83
Nerodia s. pictiventris	Florida Water Snake	85-89
Pituophis melanoleuca	Gopher Snake	45-65
Thamnophis sauritus	Ribbon Snake	80-88
Thamnophis sirtalis	Red-Sided Garter	54-84

Breeding

For the more serious herpetologists breeding is of paramount importance as wild stocks are dwindling in certain areas, and many authorities world-wide are looking to ban exports and imports of reptiles in the coming years. In order to supply the would be herpetologists of the future and guarantee the continuation of the species, herpetologists should consider breeding the animals in their care.

In order to start a breeding programme one must obviously obtain at least a pair of a particular subspecies. The authors would tend to purchase hatchlings for new breeding programmes rather than adult stock, unless purchasing from a trusted colleague. Hatchlings may take from three years upwards to mature into a healthy breeding animal, however, the wait is more than worth while.

The following methods describe how to promote breeding.

Mother and baby. Burmese Python, P.m.bivittatus.

Time of year for breeding

Most species of snake kept in captivity (and generally in their natural habitat) mate during the late or early parts of the year. Generally Boids become sexually active between the months of November through until the end of March, with most breeding activity taking place during December to February. Colubrids (in the context of this book the kingsnakes, milksnakes, watersnakes etc), do not start mating until after they emerge from hibernation. Thus Colubrids would normally be expected to mate during the months of April through to June.

How to induce a mating

Keeping male and female apart

With large Boids one can keep the male in with the female all year round without any adverse effects. With kingsnakes and milksnakes it is recommended that the pair be separated and only placed together for actual mating to take place.

CAUTION!
With those species that would normally prey on other snakes in the wild, give both snakes several meals before placing them together.

It can be quite embarrassing and expensive to find one snake being consumed by the other, as some herpetologists can testify.

Hibernation

There have been several experiments conducted throughout the world that suggest that breeding results improve if north American Colubrids are hibernated for a short period, as would happen in their natural habitat. Hibernation simply means that the temperature is dropped for a period of time with no food being given until the snake's temperature is raised to induce the end of hibernation. The authors hibernate all north American Colubrids for a minimum 8 - 10 weeks per annum at temperatures of 7°C - 13°C with no ill effects.

Spraying Water

Some species of snake that live in areas where the temperature remains more or less constant all year round rely on other factors to stimulate breeding. One factor of the natural environment is the 'rainy season'. In order to simulate this effect simply spray the animals with luke warm water quite liberally at the correct time of year and the natural instincts of the animals take over. This method is particularly useful with those species originating from tropical rain forest and seasonally dependant countries which have 'wet seasons'.

Sloughing

It has been reported that males react more sexually after the female has sloughed. This phenomena is explained by the fact that after exuviation the female's hormone (pheromone) level is higher and more readily detected by the male.

Alternatively if the sloughed skin from another male is placed within the vivaria of a previously inactive pair, you will generally find an immediate response from the original male, resulting in successful copulation.

Male to female ratio

By changing the ratio of males to females the breeding results may improve. This has been illustrated many times particularly with Colubrids and Boids, where a ratio of two males to one female triggers competition between the males in order to determine who will have the right to mate.

In the case of several species, combat dancing will result in one male's head being forced to the ground and the vanquished male then retreating to the corner of the vivarium whereupon he can be removed. With the Boids, most combat dancing does not result in injury, however, some males tend to become very aggressive to the point that serious injury will be inflicted which could lead to the death of the losing male. Thus with Boids the authors suggest that two males are closely monitored during the breeding season to prevent any outbreaks of real fighting.

Courtship Behaviour

The species of snake described in this book perform very much the same type of courtship behaviour, prior to a male copulating with the female.

A male Boid will attempt to crawl on top of the female, with his spurs held at approximately 90° to the body. The male will then proceed to scratch the female along the sides of her body in order to stimulate her into becoming receptive. In some cases the male may become so aggressive in his scratching that the female loses a scale or bleeds, although this is rarely the case.

Once the female is ready she will simply lift her tail exposing her cloaca for the male, or alternatively the male will push her tail up to allow access. Once the cloacas are aligned, copulation will commence which can last from 15 minutes to several hours.

In Colubrids, the male and female will often jerk as they sense each others presence. When the male is ready, he will attempt to crawl on top of the female and align their cloacas. In some cases the female will try and stay clear of the male, in which case the male will resort to more forceful techniques by grasping the female with his mouth on the neck region, and pushing his tail beneath hers to align the cloacas.

In their natural habitat, garter snakes and ribbon snakes congregate in their hundreds or even thousands to mate as a mass, with up to 50 males attempting to mate a single female at the same time.

Types of delivery

Oviparous and Viviparous Snakes

Snakes utilise two methods of producing young i.e. egg laying and live birth. In both cases when the female's eggs are fertilised, the eggs begin to develop within the lower part of the oviduct and often the female will stop feeding in order to avoid damaging the eggs with large meals.

In the case of an oviparous snake (pythons, ratsnakes, kingsnakes, etc) the female deposits eggs approximately 8 - 10 weeks after fertilisation. This has the advantage that the female only has to carry the eggs for a short period. Since the eggs would take up less room than living young within the oviduct, the female is capable of producing more eggs than living young. In their natural habitat the disadvantage would be that the eggs, once deposited and left unattended by the female and are vulnerable to attack from predators, as few snakes guard their clutches. The eggs take another 6 - 10 weeks at a stable temperature (32ºC for Boids, and 28ºC-30ºC for Colubrids) to hatch.

With viviparous snakes (watersnakes, boas etc), the females give birth to living young after a gestation period of 15 - 45 weeks. This too has advantages in that the female is able to maintain the young at optimum temperatures. However, it has the disadvantage in that the female becomes more restricted in movement and is therefore more open to predation.

Incubating eggs

In their natural surroundings, members of the Colubrids lay their eggs beneath logs, within piles of rotting vegetation or within moss filled crevices i.e. any area providing security and humidity at the right temperature. In captivity, the correct conditions must be replicated to encourage the females to deposit the eggs in a favourable position. This is best achieved by placing in the vivarium a plastic tub with a hole cut in the lid to allow access for the female. A large amount of damp moss or peat should be placed within the tub, alternatively damp vermiculite may be used.

Female pythons will deposit the eggs anywhere in the vivarium unless a large shallow bowl half full of damp moss or vermiculite is provided. Unlike the Colubrids, the female python will not simply lay the eggs and move on, but will gather them together within her coils to incubate and protect them.

A female Burmese Python, P.m.bivittatus, incubating her eggs.

When the female has deposited the eggs, they may be removed and placed in an incubation chamber.

CAUTION!
When removing eggs from a python, take care to secure the head of the female.

If the eggs are to be removed the female python should then be taken off the clutch and placed in a suitable container or separate vivarium whilst the eggs are collected.

The eggs are gently lifted from the female's vivarium and placed on a vermiculite and water mixture (50:50 by weight) in plastic boxes. Ensure that the eggs remain in the same position as they are collected, or the embryos may drown within its own fluid. Vermiculite is available from most D.I.Y. stores and garden centres. The eggs should not be rotated, keep them face up in the direction they are found. *Never* try to separate eggs that have stuck together as this may result in damage. The plastic box should contain several holes in the lid for ventilation. Once the eggs are placed on the vermiculite they should be lightly sprayed along with the lid and the walls of the container.

The box containing the eggs should not be placed directly on a heat pad as this will result in them baking. They should be placed on a purpose built shelf provided for the box. With the eggs in the incubation chamber and the optimum temperature selected at 32°C for boids and 28°C for Colubrids, the eggs will hatch in approximately 6 to 10 weeks. During the incubation period a bowl of water should be placed on or above the heating element to maintain a constant humidity. The correct humidity can be monitored easily without the need for a hydrometer by rubbing a finger along the inner-side of the plastic box lid within which the eggs are contained. If the humidity is correct, moisture should be present on the lid, however, if the lid is dry the lid and walls within the container should be sprayed. The humidity should be checked regularly on a daily basis from start to finish of the incubation period.

21: A clutch of kingsnake eggs prepared for artificial incubation in a margarine tub.

Python eggs being incubated within a large dustbin.

If any of the eggs discolour badly or start to smell they should be removed provided they are not stuck to any good eggs. In the case of mould growth, the slightly damaged egg should be wiped with a soft cloth and a little foot powder, as used in the treatment of foot rot or athlete's foot. This powder should not affect the eggs, only prevent the growth of mould. Any partially damaged eggs, including those that are slightly miscoloured, should be left for the duration of the incubation period as healthy youngsters may emerge.

Design of a home made incubator

In order to incubate snake eggs artificially, you must have access to an incubator. This section deals with some simple incubators that are inexpensive to construct.

Ross (1990) details three basic requirements for a python egg incubator, this principle stands for all reptile eggs.

- The construction should prevent excessive heat and humidity loss.
- There must be a consistent and reliable heat source.
- There must be a thermostat to control temperature.

At its simplist, the incubation chamber may consist of a polystyrene box (the type used to transport fish are ideal), a thermostat to regulate the temperature at 32°C, a heating pad located at the base and/or sides of the box, and a shelf on which the egg boxes are placed.

A more sophisticated incubator may be constructed from wood and perspex. The size of the incubator will depend on the species kept and the anticipated number of eggs, therefore, for one pair of snakes, the incubator can be as small as 24x18x18 inches, whereas an incubator for several species will have to be considerably larger.

The base, both side panels, rear panel and lid should be cut to size. Each piece of wood should then be treated with a heavy duty varnish and allowed to dry, alternatively damp proof paint may be used. The sides and rear panel are then secured to the base and each other using chipboard screws. With the sides and rear mounted, two pieces of 0.5^2 inch wood should be cut to length and secured to the front of the side panels, these pieces secure the front panel of perspex.

In addition several pieces of 0.5 inch wooden dowel rods should be fitted horizontally to the side panels. These rods will allow shelves to be placed at various heights within the incubator.

The lid is secured to the unit with hinges. Care should be taken to ensure an adequate seal is made.

With the basic construction complete, all panels should be lined with foam or polystyrene sheet. An electronic thermometer is then secured to a side panel and fed through to control a heat cable which is pinned to the floor and both sides of the panel.

Cut several shelves and treat with varnish or paint to prevent them deteriorating under the high humidity conditions.

As with the foam box construction, the water container should be placed directly on or above the heat element to raise the humidity, whilst the egg boxes themselves are supported on the shelves.

Live born

Those species that produce live young produce the young snake enclosed in a transparent sac which are the remains of the egg that the neonate developed in.

Under normal circumstances, the young when born will break free of their embryonic sac within a short period of time. However, if the female is under stress she may deliver the young slightly prematurely. If the young are delivered prematurely they may not be strong enough to break free of the sac and will consequently suffocate and die. Should the young not break free within a short time, tear a small hole in the side of the sac to allow the hatchlings access to air. NEVER break the cord attaching the hatchling to the last of the embryonic sac, this process must be performed by the hatchling.

The first few days

Whether hatchlings originate from the egg or directly from the mother is unimportant when looking at the first 24 - 48 hours of the hatchling's life. When the young emerge they should be placed in a container with a water dish and a damp substrate, for example wet tissue paper. This is required as most hatchlings dehydrate rapidly within the first few hours unless provided with a damp environment. After this initial stage, each hatchling may be removed and housed as required.

A Californian kingsnake, L.g.californiae, pushing its way out of the egg.

Raising Hatchlings

It is recommended that you purchase a hatchling or yearling as your first snake. Depending on the actual species selected, the snake may refuse to feed because of its specialist nature (see section on feeding) or will feed almost immediately.

Start hatchlings by placing them in a sandwich or plastic box and as they develop progressively change the box to accommodate their new size. The box should contain a piece of paper as a substrate, a water bowl and a small hide box. Hatchlings should not be overfed.

CAUTION!
Great care must be taken in handling hatchlings due to their size and delicate nature. Always support their body and do not pick them up by the neck.

Change their water every other day or as soon as it becomes soiled, as they regularly deposit their faeces in the water container. Do not place too large a water container in their box and do not fill it to the top as this will cause the water to spill over onto the substrate, possibly causing water blisters and other scale ailments.

If you are bitten by the hatchling do not jerk your hand away. Hatchlings only have very small teeth and will not puncture the skin or cause you any damage. If you pull your hand away, you could damage the hatchling's mouth.

A newly born hatchling will often refuse all food until after its first slough

Handling Snakes

Many snakes are very nervous creatures and require careful handling to prevent damage to the snake and the handler. The following points act as guidelines for the safe handling of your animals.

- When approaching a snake, hold the back of your hand towards it. Using this approach, should the snake strike it will have less of a chance of securing a hold.
- For small, non aggressive, snakes gently support the snake under the midbody whilst preventing it from gliding away with the other hand.
- For medium sized aggressive snakes hold the head between thumb and forefinger, with the index finger below the line of the jaw. The other hand should be used to support the snake at midbody.

Handling a large python. Note how the mid section of the body is supported, whilst holding the neck to prevent the animal striking out.

- For large Boids ask for the assistance of another person, to ensure that should a problem occur there is little danger.
- NEVER let a snake approach your face unless you have a firm secure hold behind its head.
- NEVER give a snake to another person unless they are experienced enough to cope or have specifically asked for the animal.
- With smaller and medium sized snakes allow the snake freedom to glide between your hands, being very cautious not to allow the snake to bite or distress anyone in close proximity.
- NEVER make a fast movement in front of the snake, as this may cause the snake to become nervous and strike out.
- NEVER remove a snake from its vivarium if another pet (cat, dog etc) is loose in the same room.
- NEVER attempt to handle your pet snake if you have not washed your hands after handling other animals, such as a pet rabbit, mice or rats.
- NEVER attempt to handle a snake after it has recently fed to allow the food to digest.

Snake

The Speckled Kingsnake *Lampropeltis getulus holbrooki.*

The colourful Mexican Milksnake *Lampropeltis triangulum annulata.*

Cornsnakes vary enormously in their colours, this is an Amelanistic Cornsnake (lacking black pigment) *Elaphe guttata guttata.*

The Snow Corn, an unusual colour morph of the Red Rat Snake.

A sub adult Western Hognose Snake *Heterodon nasicus.*

The ambition of all serious herpetologists – breeding. Here a Burmese Python incubates her clutch of eggs.

Successfully incubated, the baby Burmese Python emerges into the world.

Right: The Royal Python *Python regius* is a very popular snake, but can be a fussy feeder.

Below: The Albino 'Golden' phase of the Burmese Python *Python molurus bivittatus.*

Bottom: The head of a female African Rock Python *Python sebae.*

A young Emerald Tree Boa *Corallus caninus*.

The Columbian Boa Constrictor *Boa constrictor imperator*.

Longevity

One question that spring to mind is how long can I expect my snake to live? This is a very difficult question to answer as so many factors will affect the life span. Without question we can expect our snakes to survive far longer than a wild snake because food is readily available, there is a lack of predators and little chance of the animal being affected by extreme environmental changes.

The table below provides a guide to the life spans recorded in various collections world-wide. This table has been produced with reference to Trutnau (1986).

Longevity Records for Snakes

Scientific Name	Common Name	Age(Yrs)
Boa constrictor	Boa Constrictor	40
Loxocemus bicolor	Mexican Python	32
Python molurus molurus	Indian Python	31
Python regius	Royal Python	30
Agkistrodon contortrix mokasen	Northern Cottonmouth	29
Naja melanoleuca	Black Lipped Cobra	29
Eunectes murinus	Anaconda	28
Epicrates cenchria maurus	Rainbow boa	27
Python sebae	African Python	27
Drymarchon corais couperi	Indigo Snake	26
Crotalus atrox	Western Diamondback	25
Elaphe guttata guttata	Corn Snake	22
Lampropeltis getulus getulus	Common Kingsnake	21
Agkistrodon piscivorus	Cottonmouth	21
Lampropeltis triangulum amaura	Louisiana Milk Snake	20
Pituophis catenifer annectens	San Diego Gopher	20
Elaphe obsoleta obsoleta	Black Ratsnake	20
Lichanura trivirgata	Rosy Boa	18
Thamnophis proximus rubrilineatus	Red-striped ribbon	11

Diseases

This section looks at some of the common diseases and problems associated with keeping snakes in captivity.

We must stress that we are not encouraging home treatment of these diseases, any problem must be discussed with a veterinary surgeon before any treatment commences.

Many diseases which infect snakes are brought on by poor conditions in captivity. Many imported specimens are often riddled with disease before they arrive in this country, some importers neglect the quarantine conditions and pass the stock on to retailers as quickly as possible. This stock is then purchased by the unsuspecting customer and within a very short time dies. All stock should be carefully quarantined to ensure it does not pass on any potentially serious disease to your captive animals.

Some snakes imported from Asia have a reputation for carrying diseases which are difficult to eradicate. Some specimens, when first purchased, may be infected with mites and ticks. These are not particularly dangerous in themselves but act as vectors, carrying other diseases from one specimen to another as they suck the host's blood. The eradication of mites and ticks is not difficult.

A common occurrence is the disease known as mouth rot or necrotic stomatitis. One of the symptoms seen is the inability of the snake to close its mouth. It will then eventually refuse food and a cheesy substance may be seen inside the mouth. If mouth rot is suspected, the snake's mouth should be cleaned using a cotton bud and a 5% solution of hydrogen peroxide which is available from most pharmacists. This will remove the damaged tissue and should be followed by Betadine solution painted onto the affected area using another cotton bud. A veterinary surgeon will be able to take a swab and identify the bacteria causing the problem and provide a suitable antibiotic. It is often a Pseudomonas type of bacteria which is the culprit.

Skin infections sometimes occur in captive snakes, but these are usually brought on by keeping the specimen in poor conditions. It is very important that all specimens are kept in dry conditions. If the snake is too damp, blisters will form on the skin which will need treating. The vivarium should have a dry substrate to aid healing, for example newspaper. Damp conditions are often brought about by the lack of ventilation in the vivarium, which leads to condensation forming. This is a common problem in all glass vivaria.

Other forms of skin associated problems may include sloughing difficulties (See Ecdysis).

Some species of snake are prone to respiratory infections, which can be brought on by incorrect temperatures or a sudden temperature drop. The snake can be heard wheezing and mucous may be seen bubbling from the snake's mouth and nostrils. In some cases the snake may be seen gaping its mouth and holding its head up in the air. In all cases the specimen should be seen by a veterinary surgeon who can then suggest a course of suitable antibiotics.

Other types of disease may infect the snake's alimentary system. If a snake constantly regurgitates or produces slimy faeces it is possible that it has a micro organism infection. One that is quite common is caused by amoebic protozoa. This is extremely contagious and if left will prove fatal to the snake. It can be successfully treated using drugs, but the herpetologist must ensure the disease is not passed on to other specimens in the collection.

Many imported snakes are hosts to intestinal worms. These consist of two main types, nematodes (or roundworms) and tapeworms which can be detected in the faeces that are passed. Many intestinal diseases can be diagnosed by using microscopic investigation of passed faeces. Samples can be analysed and suitable treatment decided by your veterinary surgeon.

There are of course many other diseases which unfortunately affect snakes, but if the snake has been bred in captivity and is looked after in a conscientious manner, disease and other associated problems should not be a major handicap in the successful care and future breeding of the species in your collection.

Guide to Icons

The following section details the snakes that the authors believe are the most commonly purchased animals available through the pet trade. In each case the individual subspecies is described, paying particular attention to its requirements in captivity, its ease of maintenance and suitability for the keeper.

The table and icons detailed below describe in summary the requirements of the subspecies in as short a format as possible. This table therefore provides the reader with a fast look up table of information to quickly determine if the subspecies in question can be kept without problem.

	Denotes the country of origin from which the subspecies comes.
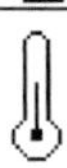	Denotes the temperature requirements of the species for both day and night, and where applicable the winter months. The most suitable forms of heating equipment using codes previously detailed are provided, with reference to the information presented in the chapter on husbandry.
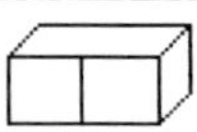	Indicates the type of vivaria and its dimensions (for both hatchlings and adults) and substrates most suitable for the subspecies with reference to the coding information
	Indicates suitable prey items to be offered to both hatchling and adult. Where necessary special notes are provided for problematic forms.
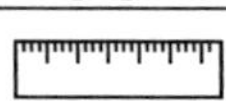	Indicates the average and maximum recorded size that adult specimens may attain.
	Indicates the suitability of the subspecies for the novice keeper, along with any special notes or warnings which should be considered before purchase.

Pythons

The Burmese Python (Python molurus bivittatus)

	Originates from Burma, Thailand, China, Hong Kong, Vietnam, Malaysia, Indonesia, north-eastern India and Indo-China.
	Spring and Summer: Day 28°C - 32°C; Night 27°C - 30°C Winter months: Day 28°C - 31°C; Night 25°C - 28°C Use heating elements code: B, E or F.
	Vivaria code: B only. Hatchling vivaria: 36 x 18 x 18 inches. Adult vivaria: 8 x 4 x 4 feet. Substrates: A (most effective), B, E or J.
	Hatchlings: Mice, small rats and pink rabbits. Adult: Large Rats, rabbits, poultry.
	Hatchlings: Average 12 - 24 inches. Adults: Average 16 - 18 feet. Maximum recorded length: 24 feet.
	Not suitable as a first time snake due to its large size. However this subspecies is generally placid and adapts well to captivity.

The Burmese Python is a large, heavy bodied constrictor found in a diverse range of habitats throughout south-east Asia from Burma across to China. It is closely related to the Indian Python (Python molurus molurus), and is often mistakenly referred to as the Indian Python when offered for sale in the local pet shops.

The usual colouring of the Burmese consists of a creamy base with large dark brown blotches bordered by yellow or dark cream bands along the back and sides. On the snake's head is a distinctive dark arrow shape extending from the neck and ending in a point above the snout. A black line is often found running through the eye on either side of the head. The belly region is a plain creamy colour.

The Burmese has for many years been a popular vivarium subject, initially imported in large numbers. However, recent advances in reptile care have led to several successful captive breeding programmes which have ensured a steady and regular supply to the pet trade. The snake's eagerness to breed in captivity has kept prices at a reasonable level and the attractive colouration has led to many herpetologists keeping this species. They are generally docile in captivity but can vary in temperament, an aggressive individual being quite a handful and relatively dangerous, particularly as they grow beyond 10feet in length. Young Burmese are often more snappy than adults and only settle as they become handled more frequently.

The Burmese Python, P.m.bivittatus. Note the difference between the adult female and the hatchling of 6 weeks.

Before purchasing a Burmese it is imperative that the would be buyer appreciates the size that these snakes attain. The hatchlings are only 12 - 24 inches long and as thick as your finger, but in three years they are likely to be 10 feet long and as thick as your leg. They have the potential to reach 16 - 18 feet and obviously a great deal of care and expertise is necessary when looking after a snake which is capable of reaching this size.

Females are mature enough to breed at lengths of 11 - 13 feet, although males which are only 8 feet have successfully fertilised females. Females deposit between 10 - 60 large oval eggs which take approximately 8 weeks to hatch at an incubation temperature of 32°C. Females will vigorously protect their eggs during the incubation period, even though many are docile at other times of the year. Whilst the eggs are within her coils, the female will be seen to jerk through muscular spasms. This causes the temperature around the eggs to rise to that preferred incubation. Hatchling Burmese will accept adult mice immediately, progressing onto rats and rabbits as they mature. Some adult Burmese also relish large poultry if available.

The African Python (Python sebae sebae)

	Found only on mainland Africa from south of the Sahara down through central Africa as far as south Africa.
	Spring and Summer: Day 28°C - 32°C; Night 27°C - 30°C Winter months: ay 28°C - 31°C; Night 25°C - 28°C Use heating elements code: B, E or F.
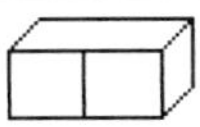	Vivaria code: B only. Hatchling vivaria: 36 x 18 x 18 inches. Adult vivaria: 8 x 4 x 4 feet. Substrates: A (most effective), B, E or J.
	Hatchlings: Mice, small rats and pink rabbits. Adult: Large Rats, rabbits, poultry.
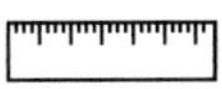	Hatchlings: Average 16 - 22 inches. Adults: Average 16 - 24 feet. Maximum recorded length: 26 feet.
	Not suitable as a first time snake due to its large size. This subspecies also has an untrustworthy disposition.

The African Python is another of the large, heavy bodied constrictors found primarily in open savannah and grasslands in the vicinity of water from south of the Sahara to central and southern Africa. The African Python is similar in appearance to that described for the Burmese, although their body shape tends to be of a smaller build being less stocky.

The usual colouration of the African Python consists of a light brown to grey brown base, with dark brown to reddish brown/black edged blotches along the back. There are smaller blotches with lighter centres, darkening towards the edges, along the sides. Unlike the head of the Burmese Python, which is rather rounded in appearance at the nasal section, the head of the African is very pointed. The head carries a dark spearhead mark, flanked on both sides by a light band from the neck region to a point above the nostrils. A black line is often found running through the eye on either side of the head. The belly region is white with black flecks.

African Python, P.s.sebae. Note the similarities with the head shape and pattern of the Burmese Python, P.m.bivittatus.

The African Python has not been imported as frequently as the Burmese due primarily to their unpredictable nature. Consequently this species is rarely bred in captivity, although captive bred specimens are more docile than their wild caught counterparts. Both authors have had limited success with this subspecies in that both have produced clutches of infertile eggs at the time of writing. Young specimens are again far more likely to show aggression than adults, although many wild caught adults will be quick to bite and cause serious lacerated wounds which require medical attention.

As for the Burmese, the prospective purchaser of an African Python should consider their unpredictable nature and their size. Hatchlings are 16 - 22 inches reaching 10 feet in approximately 4 years. They have the potential to reach 16 - 24 feet and require great expertise to avoid injury. They are quite hardy in captivity and generally easy to care for.

Females are mature enough to breed at lengths of 11 - 13 feet, males at 8 feet upwards. The females deposit between 10 - 60 eggs which are guarded during the incubation period of 8 - 12 weeks. Optimum incubation temperature is 32°C. An interesting observation is that female African Pythons do not twitch whilst incubating their eggs.

Hatchling Africans will accept adult mice immediately, progressing onto rats and as adults most frequently rabbits and large poultry as per the Burmese python.

The Royal Python (Python regius)

	Found only on mainland Africa limited in its range to countries in western and central Africa including Sierra Leone, Togo, Ghana and Senegal.
	Spring and Summer: Day 28°C - 32°C; Night 27°C - 30°C Winter months: Day 28°C - 31°C; Night 24°C - 25°C Use heating elements code: B, C, D, E or F.
	Vivaria code: A, B or C. (Note: C especially useful with problematic feeders and hatchlings). Hatchling vivaria: min: 12 x 8 x 8 inches. Adult vivaria: 36 x 18 x 18 inches. Substrates: A (most effective), B, C, E, H or J.
	Hatchlings: Mice, small gerbils, small hamsters. Adult: Mice, small rats, gerbils, hamsters (limited feeding only).
	Hatchlings: Average 8 - 14 inches. Adults: Average 3.5 - 5 feet. Maximum recorded length: 6 feet.
	Not suitable as a first time snake due to its problematic feeding habits, unless obtained from a reliable source sold as feeding.

The Royal Python is one of the smallest members of the python group and is offered for sale in most pet stores. It is a very stocky animal measuring on average only 3.5 - 5 feet, with generally a docile disposition. Found in many of the western and central countries of Africa, like the African Python it frequents water channels as they pass through forested areas and savannah.

The base colour for the Royal Python is chocolate to dark brown, interrupted by several large yellowish to greyish-white blotches over the backs and sides, normally with brown speckles in the centre of each blotch. In exceptional circumstances, this greyish-white area extends to cover most of the body area. The belly area is glossy white or cream. The head pattern is rather pointed and dark in colour with a narrow yellowish stripe running from the nose to the neck region, passing just above the eye.

Unfortunately the Royal Python has been widely imported due to their generally docile nature and small size. However, large numbers of Royals die within two years of importation due to stress and their habit of becoming anorexic. Currently very few are bred in captivity, although captive bred and captive hatched youngsters have fewer problems than wild imports.

This species should not be considered suitable as a first time purchase as a considerable amount of care must be exercised in order to maintain them correctly. Hatchlings measure 8 - 14 inches, whilst adults can reach 6 feet. Sexual maturity occurs at a length of 3.5 feet. Females deposit 4 - 9 large oval eggs, which are incubated in true python form with the female coiled around them for approximately 6 - 8 weeks.

The Royal accepts mice, small rats, gerbils, small hamsters, small guinea pigs and small poultry. This snake is so problematic in feeding that one should refer to the section covering difficult feeding behaviour to overcome the majority which prove to be troublesome. However, one must also be aware that Royal pythons fast during the winter months, and so refusing food over this period is quite natural.

The Reticulated Python (Python reticulatus)

	Found only in south-east Asia from Burma across western China including the Malay Peninsular, Borneo, Java, Sumatra, Timor, Ceram and the Philippines.
	Spring and Summer: Day 28°C - 32°C; Night 27°C - 30°C Winter months: Day 28°C - 32°C; Night 27°C - 30°C Use heating elements code: B, E or F.
	Vivaria code: B only. Hatchling vivaria: 36 x 18 x 18 inches. Adult vivaria: 10 x 4 x 4 feet. Substrates: A (most effective), B, E, H or J.
	Hatchlings: Mice, small rats, pink rabbits. Adult: Rats, rabbits, poultry (favourite item).
	Hatchlings: Average 18 - 24 inches. Adults: Average 18 - 24 feet. Maximum recorded length: 33.5 feet.
	Not suitable as a first time snake due to its untrustworthy disposition and extremely large size.

The Reticulated Python is one of the largest pythons in the world, measuring at maximum 33.5 feet in length. Although it is very long, this python tends to have a smaller build than that of the Burmese Python. It originates from most of the countries within south-east Asia, where it frequents dense, hot, humid jungle.

The base colour is yellow to dark brown, particularly around the head region. The back is covered with rhombic like blotches of yellow or yellowish-brown, bordered by black. These blotches join to form a wide zig-zag effect running the entire length of the body. The belly region is white to yellow with small black flecks. The head carries a black line from the centre of the nasals to the nape of the neck, in addition two black lines also run from the eyes to the neck. The lower jaw tends to be light in colour, either white or yellow with darker blotches and flecks interrupting each scale.

At the time of writing only two individuals were known to successfully breed this species in the United Kingdom. Captive bred individuals tend to be less aggressive than their wild counterparts, which are unpredictable and can inflict serious injury.

Snake

The prospective purchaser should consider the size and temperament of this species before purchase. Hatchlings measure 24 inches maturing to an adult size of 18 - 24 feet average, requiring an extremely large vivarium. Some expertise is required to safely maintain this species. Females deposit 15 - 100 large eggs, which the female protects. The eggs take approximately 10 - 13 weeks to hatch at an incubation temperature of 32°C. Hatchlings accept chicks, mice, small rats, hamsters and guinea pigs, progressing to large rats, rabbits, poultry and small goats when adult.

Boas

The Common Boa (Boa constrictor)

	This species has a large range, extending from Mexico in the north down to Argentina in the south.
	Spring and Summer: Day 30°C - 34°C; Night 28°C - 30°C Winter months: Day 23°C - 28°C; Night 20°C - 22°C Use heating elements code: A, B, D, E or F.
	Vivaria code: A, B or C. Hatchling vivaria: 24 x 18 x 18 inches. Adult vivaria: 4 x 2 x 3 feet. Substrates: A (most effective), B, C, E, H or J.
	Hatchlings: Mice, small rats. Adult: Mice, small rats, small rabbits.
	Hatchlings: Average 10 - 20 inches. Adults: Average 4 - 9 feet. Maximum recorded length: 14 feet.
	Generally a docile snake and one of the most suitable of the large boids for the novice keeper.

The Common Boa is the most popular of Boas available to the general public through the pet trade. It is quite a large thick bodied animal reaching an average of 9 feet in length for animals originating from Argentina and southern America, and 4 - 5 feet in length for animals originating from central America to Mexico, where they frequent jungle, savannah and cultivated fields, often in the vicinity of arid areas.

The base colour tends to vary depending on their origin, the specimens from the northern end of their range being darker than their southern cousins. In general the base colour is reddish-grey to reddish-brown with an alternating pattern of light and dark blotches running along the back. As the blotches near the tail, they become more reddish in colour, some becoming so brightly coloured as to be called 'Red-Tailed' Boas. A very fine spear like mark extends from a point between the nasals to the nape of the neck, in addition some specimens carry another bar which travels across the head in front of or between the eyes. The belly is cream to greyish with black flecks or dots.

The Boa constrictor has been imported in relatively large numbers until recently, however the trend towards successful captive breeding programmes has led to the majority of Boas on offer originating from private keepers. Hatchlings measure only 10 - 20 inches, averaging a length of 6 - 8 feet when adult. Boas produce live young, the female depositing between 15 - 60 hatchlings after a gestation of approximately 6 months.

The Common Boa is an ideal first time large boid for the novice keeper, being of a generally quiet disposition, although exceptions are found. Those animals originating from central America as imports tend to be very aggressive. Ventilation is particularly important with this species as is the higher temperature.

Hatchlings accept mice progressing to take large rats and rabbits when adult. For those individuals considering the purchase of the 'Red Tail Boas', one should beware the exceptionally high prices, especially for those animals from Surinam.

Clouded or Hogs-Island Boa (Boa constrictor nebulosa)

	This subspecies is restricted in its range to the island of Dominica.
	Spring and Summer: Day 30°C - 34°C; Night 28°C - 30°C Winter months: Day 28°C - 30°C; Night 27°C - 28°C Use heating elements code: A, B, D, E or F.
	Vivaria code: A, B or C. (C is effective with hatchlings). Hatchling vivaria: 18 x 12 x 12 inches. Adult vivaria: 48 x 18 x 18 inches. Substrates: A (most effective), B, C, E, H or J.
	Hatchlings: Mice, pink rats. Adult: Mice, small rats, pink rabbits.
	Hatchlings: Average 10 - 14 inches. Adults: Average 4 - 5 feet. Maximum recorded length: 7 feet.
	Generally a docile snake, not as large as the common boa. Suitable as a first time snake, although difficult to obtain.

This species is becoming more popular in recent times due to increased captive breeding success. The species originates from the island of Dominica where it now nears extinction, with more specimens held in captivity than on the island itself.

The base colour is whitish-grey to light tan with several small bars and/or blotches across the back. In addition the region closest to the belly scales may have flecks of yellow, blue, violet or green present giving a startling effect. The belly region is generally light cream to light tan, with small brown flecks present on some specimens.

As described, until recently this species was imported, however, due to their near extinction in their natural habitat are now under strict protection. Specimens are readily available through the pet trade as a result of intensive captive breeding programs. Hatchlings average 10 - 14 inches, reaching an average adult size of 5 feet.

When purchasing this species, one should beware their general nervous disposition, some specimens striking at every opportunity.

As with the Common Boa, adequate ventilation is required to ensure their well being.

Hatchlings take large mice, whereas adults can be offered rats and small rabbits.

The Emerald Tree Boa (Corallus caninus)

	This species is predominantly found in Brazil, Guyana and Surinam.
	Spring and Summer: Day 28°C - 30°C; Night 27°C - 28°C Winter months: Day 27°C - 28°C; Night 23°C - 27°C Use heating elements code: A, B, D, E or F.
	Vivaria code: A or B. Hatchling vivaria: 18 x 12 x 12 inches. Adult vivaria: 36 x 20 x 36 inches. Substrates: A (most effective),C or H.
	Hatchlings: Small mice, rat pups and quail. Adult: Rats, chicks and quail.
	Hatchlings: Average 10 inches. Adults: Average 3 - 4 feet. Maximum recorded length: 6 feet.
	Imported adults tend to be aggressive and resent handling. Captive bred hatchlings are more tolerant, but be careful not to damage the delicate tail on young specimens.

The Emerald Tree Boa is a brightly coloured arboreal snake found in South America. The dorsal surface is bright green with transverse white bars, whereas the ventral surface ranges in colour from a sulphur yellow to a pale cream. The young are often a brick red in appearance at birth, but gradually undergo an ontogenetic change to adult colouration during their first year.

In its natural habitat the Emerald tree Boa rarely descends from the trees, often spending long periods of time draped in a characteristic pose across the branches of its favourite resting place. It is most active at night, feeding more readily during the hours of darkness. In the wild research has shown that a large proportion of its prey consists of roosting birds which it grabs with its huge backward curving teeth. These long teeth can also cause deep wounds to any careless handler. The adults may also show displeasure by vigorously waving their tails.

In the past, specimens imported from Guyana have been available from the pet trade, they are not, however, particularly hardy, often succumbing within a few months of captivity. They frequently begin to regurgitate food and seem particularly prone to dehydration, quickly followed by death if humidity levels are too low.

More recently a few captive bred specimens have appeared on the herpetological market with successful breeding taking place in the USA and England. The captive bred young are much hardier than the wild caught specimens, eagerly feeding and showing more tolerance to lower humidity levels. Most captive bred specimens will feed readily on mice, although finicky feeders may need starting on a small quail. A useful tip to stimulate Emeralds into feeding is to place the food item in very hot water for a few seconds and then hold the prey with tongs a short distance away from the snake's heat sensitive pits. This action often stimulates a feeding response with otherwise difficult feeders.

Currently Emeralds are still an expensive species to purchase, but with increasing knowledge about the captive requirements and a greater production of captive specimens, more herpetologists are beginning to show an interest in keeping this beautiful and fascinating snake.

Kingsnakes and Milksnakes

It is best to group several subspecies under the general headings of Kingsnakes and Milksnakes, as each subspecies described require the same husbandry techniques. Although several subspecies are missing from the descriptions, they are intentionally omitted due to their difficulty in management and they are not suitable for the novice keeper.

The Kingsnakes (Lampropeltis getulus ssp)

	Found throughout the U.S.A. from east to west coast. Some subspecies extend south into Mexico.
	Spring and Summer: Day 27°C - 30°C; Night 25°C - 28°C Winter months: Day 7°C - 13°C; Night 7°C - 13°C Use heating elements code: A, C, D, E or G.
	Vivaria code: A, B or C. Hatchling vivaria: 18 x 12 x 12 inches. Adult vivaria: 40 x 18 x 18 inches. Substrates: A (most effective), B, C, E, H, I, J or K.
	Hatchlings: Pink mice, pink rats. Adult: Mice, small rats, pink rabbits.
	Hatchlings: Average 6 - 14 inches. Adults: Average 3 - 5 feet. Maximum recorded length: 6 feet.
	This group of snakes represent the most commonly kept colubrid by both the experienced and novice alike. They are ideal as a first time purchase, easily maintained and readily breed in captivity.

The kingsnakes are perhaps the most popular subspecies of snake kept by the novice and experienced herpetologist alike. They have a wide habitat ranging from desert scrub-land to forest groves and rocky hillsides, found throughout the whole of the United States of America and northern Mexico.

The Californian Kingsnake (Lampropeltis getulus californiae)
The Californian Kingsnake is the most variable patterned kingsnake, available as striped, banded or integrade forms, with base colouration being any shade of brown through to black with yellow to white bands or stripes. The Californian Kingsnake is quite robust measuring an average adult length of 3.5 - 6 feet, hatchlings measure only 8 - 12 inches. Originates from Baja California, Oregon, Utah and Arizona.

Striped phase Californian Kingsnake, L.g.californiae.

The Florida Kingsnake (Lampropeltis getulus floridana)
The Florida Kingsnake is the largest member of the kingsnake group measuring a maximum length of 72 inches, however most average 4 - 6 feet. The base colour is dark tan to dark chocolate brown with white to yellow centers which form irregular incomplete bands across the back. The neck is very thick set giving the impression that the head is attached directly to the body with no neck present. The Florida Kingsnake is restricted in its range to central and southern Florida.

The Speckled Kingsnake (Lampropeltis getulus holbrooki)

The Speckled Kingsnake is distinctive in that the base colour is generally very dark brown or most commonly black with each scale carrying a light white or yellow dot. This subspecies is relatively small with adults rarely exceeding 40 inches in length. Speckled Kingsnakes are often less docile than other kingsnakes, but are still suitable as a first time snake or as an interesting addition to a collection. The Speckled Kingsnake has a large range extending through Texas, Iowa, Illinois, Mississippi and Alabama.

The Chain Kingsnake (Lampropeltis getulus getulus)

The Chain King or Common Kingsnake is very similar in size to the Californian Kingsnake attaining an average adult length of 3.5 - 6 feet. The base colour is dark brown to black with several thin white bars crossing the back, and joining before they reach the belly area. Another of the kingsnakes with a restricted range, originating from New Jersey, Florida and Alabama.

Mexican Black Kingsnake (Lampropeltis getulus nigritus)

The Mexican Black King is perhaps the most impressive of the kingsnakes, being almost entirely black except for a few white flecks around the throat and belly regions. Typical of the kingsnakes they are stocky with a small almost unnoticeable neck, adults growing to an average of 3 - 4 feet in length. Located in the states of western Sonora and north western Sinaloa.

All of the kingsnakes are still imported although recently the importation has become unnecessary due to the large numbers produced by private collections. All kingsnakes require hibernation in order to ensure their well being and fertile eggs. The female, once provided with a container for her eggs, will deposit between 8 - 25 and leave them as previously described to incubate themselves.

The prospective purchaser should remember that all kingsnakes prey on other snakes as part of their natural food chain, therefore members of this group can only be kept in isolation with the exception of the breeding season.

All kingsnake hatchlings may be offered pink mice progressing to adult mice and small rats when adult.

The Milksnakes (Lampropeltis triangulum ssp.)

	The milksnakes have a vast range, extending from the U.S.A. in the north, south through Mexico to South America.
	Spring and Summer: Day 27°C - 30°C; Night 25°C - 28°C Winter months: Day 7°C - 13°C; Night 7°C - 13°C Use heating elements code: A, C, D, E or G.
	Vivaria code: A, B or C. (Type C is ideal for problematic feeders). Hatchling vivaria: min: 12 x 10 x 10 inches. Adult vivaria: 36 x 18 x 18 inches. Substrates: A (most effective), B, C, E, H, I, J or K.
	Hatchlings: Pink mice, pink rats. Adult: Mice, small rats, pink rabbits.
	Hatchlings: Average 6 - 14 inches. Adults: Average 2 - 4 feet. Maximum recorded length: 6 feet.
	This group is generally more difficult to keep than the kingsnakes, due to feeding problems with some hatchlings. The milksnakes described represent the easier subspecies to maintain.

The milksnakes are perhaps the most colourful of those species available to the novice keeper with several suitable forms available. Milksnakes extend from central south America to the lower parts of the U.S.A. They range in size from 3 - 6 feet depending on the subspecies, most being characterised by their bands of red, black and white.

The Mexican Milksnake (Lampropeltis triangulum annulata)

The Mexican Milksnake is one of the most commonly offered milksnake in the pet trade, having many small red bands interrupted by bands of white or yellow bordered by thin bands of black on either side. The belly region is generally black or with few bands crossing over. This subspecies is very hardy in captivity and grows to a reasonable size of only 3 - 4 feet when adult and of very stocky proportions. This subspecies is restricted to southern Texas and Mexico.

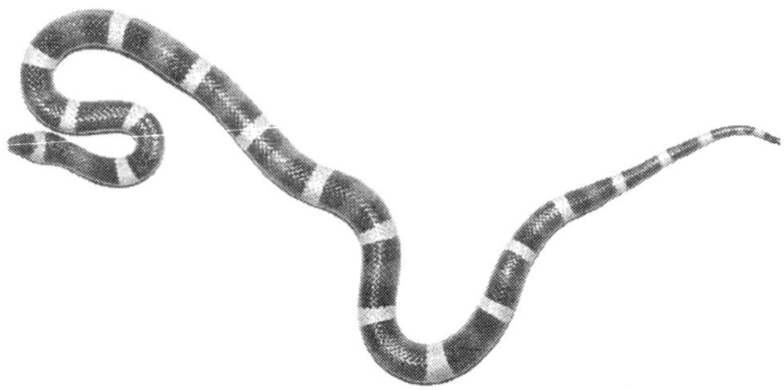

A typical milksnake, L.t. campbelli.

The Sinaloan Milksnake (Lampropeltis triangulum sinaloae)
The Sinaloan Milksnake is another of the popular forms of milksnakes available, having very wide red bands with narrow white bands bordered by black on either side. The Sinaloan milksnake although being similar in body length to the Mexican Milksnake, is less stocky being of a smaller build. Found only in a small region in western Mexico.

The Honduran Milksnake (Lampropeltis triangulum hondurensis)
The Honduran Milksnake is one of the largest members of the milksnake group attaining an average adult size of 4 - 6.5 feet. It is thick bodied and displays the typical bands of the group. This subspecies is available as three colour morphs, normal; tangerine or orange. The normal phase has red bands with the red scales tipped by black. These red bands are again interrupted by white bands bordered by black, often with the white scales flecked with black. The tangerine and orange phases have all scales tangerine or orange with slightly darker pigment available where the normal borders of the tri-coloured bands would occur.

CAUTION!
Imported Honduran milksnakes are generally aggressive and will take every opportunity to bite and defecate when handled.

The Pueblan Milksnake (Lampropeltis triangulum campbelli)
The Pueblan Milksnake is fast becoming a popular subject, sold mainly through specialist outlets rather than the average pet store. It has a distinctive black snout often flecked with white and a large broad white band on the anterior section of the head and extending back to several scales width on the neck and back. Thereafter the body has the typical tri colour markings of the milksnakes. Adults reach an average of 2.5 - 3 feet, hatchlings emerging from the egg only 6 - 8 inches. Some hatchlings can be finicky with their first meals and may require scenting of their pinkies with frogs and lizards. The Pueblan Milksnake is another of the milksnakes with a very limited range, found only in a remote mountain range in south central Mexico.

The comments provided for the kingsnakes will ensure the correct husbandry of the milksnakes.

Ratsnakes

Like the milksnakes and kingsnakes, ratsnakes are another popular exhibit in the local pet stores, generally available as corn snakes or red ratsnake, black, grey and yellow ratsnakes. Ratsnakes are very easy to maintain in captivity being hardy and long lived. Ratsnakes are found throughout the world, however, only the north American forms should be considered as the imported Asian varieties available are generally aggressive, difficult to keep and full of parasites.

The Corn Snake (Elaphe guttata guttata)

	Found throughout central and eastern U.S.A. particularly around farms in Jersey, Florida and Louisiana.
	Spring and Summer: Day 27°C - 30°C; Night 25°C - 28°C Winter months: Day 7°C - 13°C; Night 7°C - 13°C Use heating elements code: A, C, D, E or G.
	Vivaria code: A, B or C. Hatchling vivaria: 18 x 12 x 12 inches. Adult vivaria: 36 x 18 x 18 inches. Substrates: A (most effective), B, C, E, H, I, J or K.
	Hatchlings: Pink mice, pink rats. Adult: Mice, small rats, pink rabbits.
	Hatchlings: Average 6 - 10 inches. Adults: Average 4 - 6 feet. Maximum recorded length: 6 feet.
	The Corn Snake is probably the most suitable novice snake, being of a gentle disposition, breeding readily in captivity and living many years.

The Corn snake, or Red Ratsnake, like other ratsnakes is long and slender with large round eyes and thin neck. The base colour of the body varies from rustic red to orange with red to brown blotches bordered by black and often white. Smaller blotches are present on the sides, with larger blotches restricted to the back. The head often carries a 'V' shaped mark from between the eyes to the nape of the neck. The belly tends to be white, sometimes cream, mottled with black flecks.

The Corn Snake, E.g.guttata, the friendliest and most commonly kept rat snake.

In addition to the normal variation described, the corn snake has the honour of being the most varied in colour, with some 20 or so colour morphs available. Only the most common forms are briefly described below:

'Snow' corn. Has a base colour of cream to greyish-white with slightly yellowish blotches.

Amelanistic corn. A corn snake lacking all black pigment and so is brightly coloured in red and orange.

Albino corn. Generally pink in colour with slightly darker shades on the blotches.

Anerythristic corn. Lacks all of the red pigment and, therfore, is mainly grey, black and white in clouration, closely resembling the Grey Ratsnake.

Adults reach an average size of 4 - 6 feet, depending on origin. Hatchlings measure 6 - 10 inches. Females mature at approximately 24 inches, depositing 8 - 26 small oval eggs, which hatch after 8 weeks to produce hatchlings of 6 - 10 inches in length. Like other hatchling snakes, they may bite during the first few weeks, but most calm to be very docile when adult.

The corn snake is a favourite amongst experienced and novice keepers alike, due to their good disposition and ease of maintenance. This is the ideal first snake to start a breeding programme as many will breed successfully, even without hibernation.

Grey Ratsnake (Elaphe obsoleta spiloides)

	Found throughout central and eastern U.S.A in the states of Illinois Indiana, Mississippi and Georgia.
	Spring and Summer: Day 27°C - 30°C; Night 25°C - 28°C Winter months: Day 7°C - 13°C; Night 7°C - 13°C Use heating elements code: A, C, D, E or G.
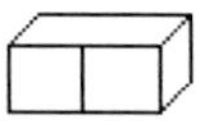	Vivaria code: A, B or C. Hatchling vivaria: 18 x 12 x 12 inches. Adult vivaria: 36 x 18 x 18 inches. Substrates: A (most effective), B, C, E, H, I, J or K.
	Hatchlings: Pink mice, pink rats. Adult: Mice, small rats, pink rabbits.
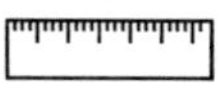	Hatchlings: Average 12 - 16 inches. Adults: Average 4.5 - 6 feet. Maximum recorded length: 6.5 feet.
	This subspecies is another of the prefered ratsnakes suitable for the novice keeper.

The second of the more docile ratsnakes, the Grey Ratsnake is another colourful addition. This subspecies is slightly larger built than the corn snake having a generally larger girth for its 6 feet length.

The base colour is whitish-grey to oak brown with rectangular type blotches running along the centre of the back. The blotches are dark grey to dark brown, with small dark brown flecks along the sides of the body. The head carries a dark bar from the neck just above the jaw angle through the eye.

The grey ratsnake is also called the Oak Ratsnake due to its excellent climbing skills and the fact that it often frequents oak trees in search of nesting birds and their eggs.

Females, like the Corn snake, mature at approximately 24 inches. However, one should, if possible, prevent females copulating with males until they reach about 48 inches. This prevents the small female becoming 'egg-bound' if the eggs are too large for her to pass. Females deposit approximately 9 - 16 eggs, which produce hatchlings of 12-16 inches in length.

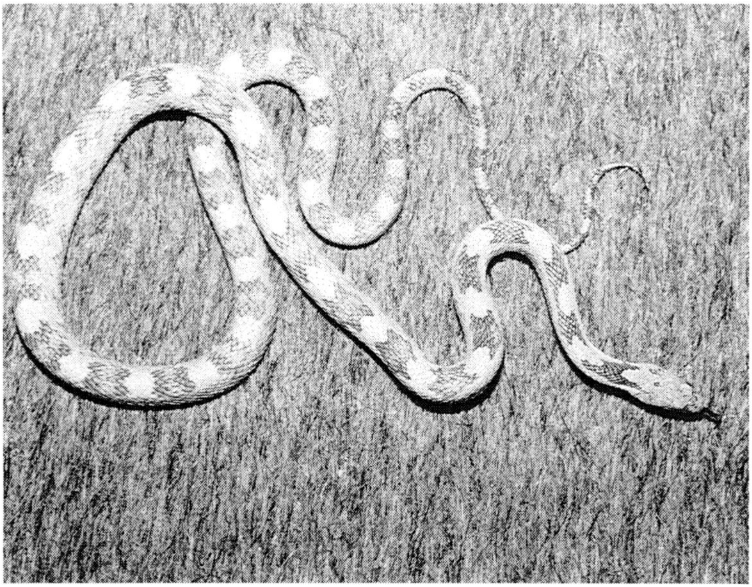

The Grey Ratsnake, E.o.spiloides, another of the docile ratsnakes favoured by the hobbyist.

Black Ratsnake (Elaphe obsoleta obsoleta)

	North eastern U.S.A. - Wisconsin, Oklahoma, Louisiana and Georgia.
	Spring and Summer: Day 27°C - 30°C; Night 25°C - 28°C Winter months: Day 7°C - 13°C; Night 7°C - 13°C Use heating elements code: A, C, D, E or G.
	Vivaria code: A, B or C. Hatchling vivaria: 18 x 12 x 12 inches. Adult vivaria: 36 x 18 x 18 inches. Substrates: A (most effective), B, C, E, H, I, J or K.
	Hatchlings: Pink mice, pink rats. Adult: Mice, small rats, pink rabbits.
	Hatchlings: Average 10 - 15 inches. Adults: Average 6 - 7 feet. Maximum recorded length: 9 feet.
	Although often found in pet stores, the Black Ratsnake will readily bite if handled, very few being docile. A suitable snake for more experienced keepers.

The Black ratsnake has a slightly less predictable nature than the previous ratsnakes described. Adults can reach almost 9 feet, however, the average length is 6 - 7 feet.

The base colour is always dark brown to black with slightly darker blotches along the back and sides. Hatchling black ratsnakes resemble the grey ratsnake in appearance being more greyish in colour with darker triangular shaped blotches.

CAUTION!

This species can be unpredictable and will deliver a painful bite if given the opportunity.

As with the Grey Ratsnake, females should only be considered suitable for breeding when they mature to 48 inches length. The female will deposit between 6 - 18 large oval eggs, which will produce hatchlings of 10 - 15 inches in length.

Yellow Ratsnake (Elaphe obsoleta quadravittata)

	Found throughout central and eastern U.S.A. including Florida Georgia, North and South Carolina.
	Spring and Summer: Day 27°C - 30°C; Night 25°C - 28°C Winter months: Day 7°C - 13°C; Night 7°C - 13°C Use heating elements code: A, C, D, E or G.
	Vivaria code: A, B or C. Hatchling vivaria: 18 x 12 x 12 inches. Adult vivaria: 36 x 18 x 18 inches. Substrates: A (most effective), B, C, E, H, I, J or K.
	Hatchlings: Pink mice, pink rats. Adult: Mice, small rats, pink rabbits.
	Hatchlings: Average 8 - 12 inches. Adults: Average 6 - 7 feet. Maximum recorded length: 8 feet.
	Another of the ratsnakes not suited to the novice keeper due to their disposition and willingness to bite.

CAUTION!
The Yellow Ratsnake is often found on display in pet stores, usually as hatchlings. The yellow ratsnake although regularly bred in captivity, is often an aggressive snake.

Yellow ratsnakes, as the name suggests, are mainly yellow being brighter when hatchling size. Adults tend to have a series of darker brown to reddish brown thin stripes running the length of the body from the nape of the neck. The yellow scales of the adults tend to darken with age to more of a tan colour.

Like the Grey Ratsnake, the Yellow Ratsnake will often climb trees in search of eggs and fledgling birds.

The prospective purchaser should consider the size of the ratsnake to be purchased and the actual subspecies, where possible avoiding the more aggressive forms if a first time purchase.

Ratsnakes will breed quite easily in captivity with females depositing between 10 - 40 small oval eggs. All ratsnakes should be hibernated during the winter months at temperatures of 7°C - 13°C.

Hatchlings measuring 8 - 12 inches in length will accept pink mice moving onto adult mice and small rats.

The Yellow Ratsnake, E.o.quadravittata, an aggressive and untrustworthy ratsnake, although exceptions can be found.

Bull and Pine Snakes

The Bull, Gopher and Pine Snakes are all very large, robust snakes. They come from Canada in the north, to Mexico in the south of the Americas. All are evenly tempered, often striking with their mouths' closed, although when they do bite, it is very painful due to the enormous strength of their jaws. All make suitable first time purchases for the novice and are represented in both novice and experienced herpetologist collections.

The Northern Pine Snake (Pituophis m. melanoleucus)

	U.S.A. in the states of New Jersey, Alabama, Virginia, Kentucky and Tennessee.
	Spring and Summer: Day 27°C - 30°C; Night 25°C - 28°C Winter months: Day 7°C - 13°C; Night 7°C - 13°C Use heating elements code: A, C, D, E or G.
	Vivaria code: A or B. Hatchling vivaria: 18 x 12 x 12 inches. Adult vivaria: 48 x 18 x 18 inches. Substrates: A (most effective), B, C, E, H, I, or J.
	Hatchlings: Small mice, pink rats. Adult: Mice, small rats, pink rabbits.
	Hatchlings: Average 10 - 18 inches. Adults: Average 4 - 5 feet. Maximum recorded length: 6.5 feet.
	The Northern Pine is the most commonly kept of the pine snakes, being handsomely marked and easily cared for. Wild imports strike repeatedly, but calm over a lengthy period of time.

The Northern Pine is very robust and typical of the family. The base colour is often white to cream with large dark brown to black blotches along the back. Although this subspecies has a reputation for being aggressive it seldom strikes with the intention of biting, but rather strikes with the mouth closed.

As with other north American Colubrids this subspecies must be hibernated during the winter months in order to guarantee breeding success. Females deposit 3 - 18 large oval eggs, which hatch to produce hatchlings of 10 - 18 inches in length. Captive bred individuals are less likely to strike than their wild colleagues.

The Black Pine Snake (Pituophis m. lodingi)

	U.S.A. - from private breeders.
	Spring and Summer: Day 27°C - 30°C; Night 25°C - 28°C Winter months: Day 7°C - 13°C; Night 7°C - 13°C Use heating elements code: A, C, D, E or G.
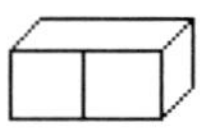	Vivaria code: A or B. Hatchling vivaria: 18 x 12 x 12 inches. Adult vivaria: 48 x 18 x 18 inches. Substrates: A (most effective), B, C, E, H, I, or J.
	Hatchlings: Small mice, pink rats. Adult: Mice, small rats, pink rabbits.
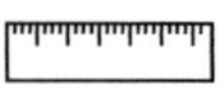	Hatchlings: Average 10 - 18 inches. Adults: Average 4 - 5 feet. Maximum recorded length: 6.5 feet.
	The Black Pine is another of the common Pine Snakes regularly sold through the pet trade. Like the Northern Pine, this subspecies is suitable for the novice keeper, although some individuals will attempt to strike when young.

The Black Pine Snake is fast becoming one of the most sought after snakes due to its dark brown to black pattern, often without any white flecks. This dark colouration gives it the appearance of a much larger, more aggressive, snake although in reality it tends to be less aggressive than its cousin the Northern Pine.

Females deposit 5 - 15 large oval eggs, which hatch to produce hatchlings of 10 - 18 inches in length. Captive bred examples are generally more expensive than other forms, but are often sought after.

The Bull Snake (Pituophis m. sayi)

	From Canada down through the central states of the U.S.A. to Mexico.
	Spring and Summer: Day 27°C - 30°C; Night 25°C - 28°C Winter months: Day 7°C - 13°C; Night 7°C - 13°C Use heating elements code: A, C, D, E or G.
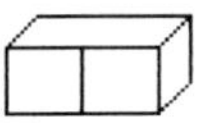	Vivaria code: A or B. Hatchling vivaria: 18 x 12 x 12 inches. Adult vivaria: 48 x 18 x 18 inches. Substrates: A (most effective), B, C, E, H, I or J.
	Hatchlings: Small mice, pink rats. Adult: Mice, small rats, pink rabbits.
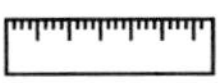	Hatchlings: Average 10 - 16 inches. Adults: Average 4 - 5 feet. Maximum recorded length: 5.25 feet.
	The Bull Snake is suitable as a first time snake due to its relatively calm disposition when raised in captivity. Some can be aggressive but, like the Pine snakes, tend to strike with their mouth closed.

The Bull Snake is slightly stouter than its cousin, the Pine Snake, with a larger head and neck region. The base colour is often cream to yellow, although in exceptional cases, some may be orange to tan. The back and sides carry large tan to chocolate brown saddles and bars running the whole length of the body.

Females deposit 3 - 24 large oval eggs, which hatch to produce hatchlings of 10 - 16 inches in length. Captive bred examples are generally more docile than wild imports.

Other Colubrids

Western Hognose (Heterodon nasicus nasicus)

	New Mexico, Texas, western Oklahoma as far north as southern Canada.
	Spring and Summer: Day 27°C - 30°C; Night 25°C - 28°C Winter months: Day 7°C - 13°C; Night 7°C - 13°C Use heating elements code: A, C, D, E or G.
	Vivaria code: A, B or C. Hatchling vivaria: 12 x 10 x 10 inches. Adult vivaria: 24 x 18 x 18 inches. Substrates: A (most effective), C, E, G, H, I, J or K.
	Hatchlings: Pink mice. Adult: Mice, pink rats.
	Hatchlings: Average 6 - 7 inches. Adults: Average 15 - 25 inches. Maximum recorded length: 32 inches.
	The Western Hognose is a favourite amongst novice and experienced keepers alike, however some caution must be exercised due to the toxicity of their saliva.

The Western Hognose is a relatively small member of the Colubrid family and was first described in 1852. It originates from a large area within northern America, extending from southern Canada throughout central U.S.A. to eastern Mexico.

The most distinctive feature of this attractive little snake is the sharply upturned and pointed snout, the overall body shape is robust and chunky with strongly keeled scales on the body. Along the back and sides there is a row of broad, dark blotches over a sandy, buff background colouration. Smaller brown patches occur on the lower sides, with the belly being predominantly black in colouration. A dark chocolate double band runs across the forehead, with this band connecting the eye to the corner of the mouth. The adults vary in size, but are normally between 15 - 25 inches in length, although specimens up to 30 inches are found on occasions. When the young hatch they are between 6 and 7 inches in length.

The Western Hognose, Heterodon nasicus nasicus, an attractive and unique snake.

These snakes have unusual teeth, in fact their name, heterodon, means different or multiple toothed. This name is given due to the unusual large rear teeth of this particular species. When in a resting position the large rear fangs lie backwards but on biting their prey these rear fangs are extended forwards and are used to inject a mildly toxic saliva.

When harassed by a predator the Western Hognose puff themselves up, spreading their neck into a hood and hiss surprisingly loudly for a small snake. Most textbooks state that a Hognose will not bite humans and if they do the saliva injected is not toxic to man. This theory may have to be modified as both authors are aware of instances where a bite has resulted in serious pain and, in one instance, a bite victim was kept in hospital overnight as a precaution. Do not be put off by these isolated incidents, the Western Hognose is generally an inoffensive specimen and is only known to bite on very rare occasions.

The Western Hognose is another species rarely imported due to the success of captive breeding programs. Hatchlings will take pink mice although their feeding may be spasmodic during the first year. Some specimens will eat readily for several months and then refuse food for many weeks before starting to feed once more.

Females produce 9 - 21 small oval eggs, similar in appearance to ratsnake eggs. The eggs take approximately 7 - 9 weeks to incubate, resulting in the relatively small hatchling size. Some youngsters can be problematic in their feeding habits, which is overcome normally by rubbing the prey item to be offered on a frog.

Youngsters are best maintained in clear plastic tubs during their first year, progressing to larger vivaria as shown in the chart.

The Water Snakes (Nerodia ssp)

U.S.A.
Spring and Summer: Day 24°C - 30°C; Night 21°C - 23°C Winter months: Day 7°C - 13-°C; Night 7°C - 13°C Use heating elements code: A, D, E or G.
Vivaria code: A or B. Hatchling vivaria: 12 x 10 x 10 inches. Adult vivaria: 36 x 18 x 18 inches. Substrates: A, C, D, G, H.
Hatchlings: Pink mice, small fish (Guppies, goldfish, lancefish, garter grub). Adult: Mice, pink rats, lancefish, goldfish, garter grub. Caution: Meals should be offered with vitamin supplements.
Hatchlings: Average 6 - 7 inches. Adults: Average 3 - 4 feet. Maximum recorded length: 6 feet.
Water snakes tend to be slightly aggressive, but are a favourite amongst many herpetologists. Suitable for the novice.

The Water snakes on display in most pet stores originate from the U.S.A. particularly from the Florida peninsula. They are heavily built, being very broad for a short body length, with rough keeled scales.

The Florida Watersnake (Nerodia fasciata pictiventris)

The Florida Watersnake is the most commonly found subspecies of watersnake in the pet stores. It is a medium sized snake of average length 3 - 4 feet, in exceptional cases up to 6 feet. The body is made up of a series of bands starting with a light band of green to reddish-tan, alternating with a darker band of dark green to dark brown. A black bar extends from the angle of the jaw to the rear of the eye. The belly is generally whitish to creamy yellow.

Although they are often aggressive, the Florida Watersnake is very easily maintained in captivity. Provided they are hibernated, the Florida Watersnake will reproduce readily, with females depositing approximately 20 - 40 young during the months of June and July.

The Banded Watersnake (Nerodia fasciata fasciata)

The Banded Watersnake is almost identical in appearance to the Florida Watersnake, having bands of a dark green to brown, alternating with even darker bands of green to near black. The belly area is whitish to creamy yellow. However, unlike the Florida Watersnake, it is chequered with rectangular dark blotches.

The Banded Watersnake has the same reproductive cycle as described for the Florida Watersnake, with females depositing between 20- 50 live young during the months of June through until August.

Most Water snakes are extremely bad tempered and will bite savagely when presented with an opportunity. They rarely settle down if wild caught, although captive bred individuals are less likely to bite.

Hatchlings will accept pink mice if a fish is rubbed against the pinky. Hatchlings raised on pinkies will grow to accept mice without the use of odour manipulation. Problem feeders should be presented with fresh water fish or par boiled whitebait, each fish being dipped into vitamin supplements. Should fish be offered without supplements or not having been boiled for at least 30 seconds, the fish will contain the enzyme thiaminase, which effectively destroys Vitamin B1, resulting in the animal's loss of balance, and in the worst case death.

Water snakes reproduce readily in captivity producing up to 50 live young during the months of June - August. They should be hibernated during the winter months.

The prospective purchaser should only consider purchasing newly hatched or captive raised individuals as a first time snake.

The Garter Snakes and Ribbon Snakes (Thamnophis ssp)

	U.S.A.
	Spring and Summer: Day 27°C - 30°C; Night 25°C - 28°C Winter months: Day 7°C - 13°C; Night 7°C - 13°C Use heating elements code: A, D, E or G.
	Vivaria code: A or B. Hatchling vivaria: 12 x 10 x 10 inches. Adult vivaria: 36 x 18 x 18 inches. Substrates: A, C, D, G, H.
	Hatchlings: Pink mice, small fish (Guppies, goldfish, lancefish, Garter grub). Adult: Mice, pink rats, lancefish, goldfish, garter grub. Caution: Meals should be offered covered with vitamin supplements.
	Hatchlings: Average 6 - 10 inches. Adults: Average 3 - 4 feet. Maximum recorded length: 5 feet.
	Although some may be aggressive, Garter snakes are a favourite amongst many herpetologists. Suitable for the novice.

The Ribbon and Garter snakes are perhaps the more nervous of those species purchased through the pet trade. These subspecies originate from most of the U.S.A. Again it is the eastern forms that are more frequently seen in the U.K. The distinguishing features between the Garter and Ribbon snakes are that Ribbon snakes tend to have heads that are well set off from the body in the adult form. Also Ribbon snakes are very slimly built, whereas Garter snakes are more heavily built.

Eastern Ribbon Snake (Thamnophis sauritus sauritus)

The Eastern Ribbon Snake is typical of the Ribbon snakes in being extremely long and thin. This subspecies has a base colour of tan to light brown interrupted on the back and sides by three bright yellow to cream stripes along the length of the body. The belly colour ranges from cream to light tan.

Eastern Garter Snake (Thamnophis sirtalis sirtalis)

The Eastern Garter Snake is the most popular of the Garter snakes in many collections. Typically the Eastern Garter averages 3 - 4 feet when adult. This subspecies has a base colour of tan to dark chocolate brown with three stripes of yellow to gold running along the length of the body. In some individuals small flecks of red, orange or blue may be present. The belly is whitish-grey to bluish-green with a small black spot on each side.

The Bluestripe Garter Snake (Thamnophis sirtalis similis)

The Bluestripe Garter Snake is similarly proportioned to the Eastern Garter Snake. The base colour tends to be dark brown to near black with the stripe on the centre of the back being a dull tan to darker brown, and the stripes on the sides of the body being light blue. The belly area is again whitish-grey to bluish in colouration.

Unfortunately Ribbon and Garter snakes continue to be imported in large numbers each year, many not surviving the first 12 months in captivity due to their nervous disposition. The vast majority of imported animals are males that are collected in their hundreds as they attempt to mate with single females shortly after emerging from hibernation.

These species feed primarily on amphibians and fish, although some can be persuaded to accept pink mice and furries through the use of odour manipulation. As with Water snakes, if feeding fish one should sprinkle vitamin supplements onto each meal to ensure the correct balance of vitamins and avoiding the results of vitamin B1 deficiency. To avoid using vitamins sprinkled on fish, one should try and purchase Garter-Grub as previously described, which has been specially formulated to meet the needs of this species.

Garter and Ribbon snakes can be kept in large colonies without ill effect. Care should be taken when feeding as some specimens will go into a feeding frenzy biting everything that moves. A small water container must be provided for them to soak. Care must be exercised to prevent spillage and excessive water remaining, as this species is prone to water blisters.

Wild imports should be given several hide boxes to provide security and help them adjust, although captive bred or long term captives are indifferent to a hide box.

The Chequered Garter snake, one of the most attractive of the family.

The House Snakes (Lamprophis ssp)

	Central and southern Africa.
	Spring and Summer: Day 27°C - 30°C; Night 25°C - 28°C Winter months: Day 13°C - 16°C; Night 13°C - 16°C Use heating elements code: A, D, E or G.
	Vivaria code: A, B or C. Hatchling vivaria: 12 x 10 x 10 inches. Adult vivaria: 24 x 18 x 18 inches. Substrates: A, B, C, E, H, I, J or K.
	Hatchlings: Pink mice, small fish. Adult: Mice, pink rats.
	Hatchlings: Average 7 - 10 inches. Adults: Average 2 - 3 feet. Maximum recorded length: 3.25 feet.
	House snakes are fast becoming one of the most popular novice snake being easily maintained in captivity and regularly bred by herpetologists.

House snakes are one of the hardiest of snakes available, able to tolerate extremes of temperature, and habitat.

All House snakes originate from Africa, particularly the southern countries. All are long and slender with shiny smooth scales. Although hatchlings can sometimes be troublesome feeders, they mature into attractive adults, willingly breeding in captivity. It is not uncommon for females to double or even triple clutch in a single year, producing between 2 - 15 small oval eggs.

The Brown House Snake (Lamprophis fuliginosus), is the most commonly sold house snake. It is uniformly tan to brown in colour, except for a cream belly and two thin cream lines running from the snout, through each eye to just behind the head.

Appendix A: Glossary of Terms

Aestivation:	To pass the summer or dry season in a dormant condition.
Anal plate:	This scale marks the beginning of the tail section and is also the scale covering the cloaca entrance.
Arboreal:	Spends much of its time in trees and high bushes.
Boids:	A family of snakes comprising the pythons and boas.
Cloaca:	A chamber into which the intestinal, urinary and reproductive canals meet and empty.
Colubrid:	A member of the largest world-wide family of snakes.
Constrictor:	A snake that kills its prey by wrapping its body around the victim thus preventing it from recovering its breath and killing it through suffocation.
Crepuscular:	Active at sunset or sunrise.
Diurnal:	Means that the snake is active during the day.
Dorsal:	Upper surface or back of the snake.
Ecdysis:	To slough or shed skin.
Ectoparasite:	External parasites like mites and ticks.
Epidermis:	Refers to the layer of skin or scales.
Exuviation:	Sloughing or the shedding of old skin.
Faeces:	The waste material excreted from the cloaca of the snake.
Gestation:	The period of time that it takes the young to develop within the female prior to birth.
Gravid:	Used to describe a pregnant snake.
Hemipene:	The Y shaped penis of snakes and lizards.
Herpetology:	The study of reptiles and amphibians.

Husbandry:	The care of wild or captive animals.
Keeled:	Hard rough scales particularly common to the Water Snakes and Ribbon and Garter Snakes.
Nasals:	The scales in which the snakes nostrils are located.
Neonate:	A new-born snake.
Nocturnal:	Snakes that are active by night.
Ophiophagous:	Snakes that take other snakes as part of their natural diet.
Oviparous:	Snakes that are egg-layers.
Ovoviviparous:	Snakes that are live-bearing.
Ovulation:	The release of mature ova in female snakes.
Pheromone:	Referring to the female's hormone level from their sex gland within the tail.
Photo-period:	The ratio of light to dark to which an animal is exposed.
Pinkies:	New-born mice or baby mice lacking hair.
Poikilothermic:	Literally translates as various heat, i.e. dependant on external factors for bodily heat.
Sloughing:	A term used to describe the shedding of a skin.
Subcaudal:	Those scales located on a snake after the cloaca to the tip of the tail on the ventral side.
Substrate:	Vivarium floor covering.
Thermoregulate:	To control temperature.
Vent:	See cloaca.
Ventral:	The belly area of a snake.

Appendix B: Herpetological Societies

Herpetological Societies offer both the amateur and professional herpetologists a means of contact with fellow keepers, advice and expertise on husbandry techniques. In addition most publish newsletters and journals with articles of interest. We would therefore advise that one joins at least one society in order to develop one's knowledge more quickly. Several types of society exist from the specialist, set up to study particular species, through to the local and international general herpetological societies which were formed to cater for any herpetological interest. The list below represent the main groups in the United Kingdom.

Milton Keynes Herpetological Society
15 Esk Way
Bletchley
Milton Keynes
Bucks.
MK3 7PW.

International Herpetological Society
Secretary/Treasurer
27 St. Thomas Close
Dartmouth Avenue
Walsall
West Midlands
WS3 1SZ.

British Herpetological Society
Headquarters
Zoological Society of London
Regent's Park
London, NW1 4RY.

Bibliography

Ashton, R.E. Jr., and P.S. Ashton. 1981. Handbook of reptiles and Amphibians of Florida, Part 1, The snakes. Windward Publishing Inc.

Behler, J.L. 1979. The Audubon Society field guide to North American reptiles and Amphibians. Alfred A. Knopf.

Cox, M.J. 1991. The Snakes of Thailand and Their Husbandry. Krieger Publishing Company.

Gotch, A.F. 1972. Reptiles: Their Latin names explained. Blandford Press, Dorset.

Klauber, L.M. 1972. Rattlesnakes: Their habits, life histories, and influence on mankind. 2nd Edition. Berkeley and Los Angeles: University of California Press. 2 vols.

Mattison, C. 1988. Keeping and breeding Snakes. Blandford Press.

Mehrtens, J.M. 1987. Living snakes of the World, in colour. Sterling Publishing Co.

Pomfret, B. 1989 - 1994. Various articles from the Milton Keynes Herpetological Society Newsletters.

Protheroe, J.P. 1989 - 1994. Various articles from the Milton Keynes Herpetological Society Newsletters.

Protheroe, J.P. 1990. Snakes: A Guide to their husbandry before and after purchase. Milton Keynes Herpetological Society.

Protheroe, J.P. 1990. Subcaudal Count Ranges. Milton Keynes Herpetological Society.

Ross, R.A., and G. Marzec. 1990. The Reproductive Husbandry of Pythons and Boas. The Institute for Herpetological Research.

Stebbins, R.C. 1966. A field guide to Western Reptiles and Amphibians. Houghton Mifflin Company, Boston.

Trutnau, L. 1986. Nonvenomous Snakes. Barron's.

Index